ADVANCES IN

FOOD AND NUTRITION RESEARCH

VOLUME 41

Starch

Basic Science to Biotechnology

ADVANCES IN
FOOD AND NUTRITION RESEARCH

VOLUME 41

Starch

Basic Science to Biotechnology

Edited by

MIRTA NOEMI SIVAK
AND
JACK PREISS
Department of Biochemistry
Michigan State University
East Lansing, Michigan

ACADEMIC PRESS
San Diego London Boston New York Sydney Tokyo Toronto

This book is printed on acid-free paper. ∞

Academic Press
a division of Harcourt Brace & Company
525 B Street, Suite 1900, San Diego, California 92101-4495, USA
http://www.apnet.com

Academic Press Limited
24-28 Oval Road, London NW1 7DX, UK
http://www.hbuk.co.uk/ap/

International Standard Book Number: 0-12-016441-8

PRINTED IN THE UNITED STATES OF AMERICA
98 99 00 01 02 03 QW 9 8 7 6 5 4 3 2 1

Dedicated to the memory of
Carlos E. Cardini and Luis F. Leloir, pioneers.

CONTENTS

PREFACE xiii

Occurrence of Starch

I. Introduction 1
II. Seeds 1
III. Storage Roots and Tubers 3
IV. Starch in the Gravitational Response of Roots and Stems 3
V. Leaves 4
VI. Green Algae 4
VII. Other Reserve Polysaccharides 5
VIII. Experimental Systems in the Study of Starch Metabolism 6
Further Readings 12

Physicochemical Structure of the Starch Granule

I. The Starch Granule 13
II. Amylose and Amylopectin 13
III. Molecular Orientation in the Granule 27
IV. Methodology and Nomenclature Used in Starch Analysis 29
V. Other Constituents of the Starch Granule 30
VI. Lipids 30

VII. Phosphorus 30
VIII. Proteins 31
Further Readings 32

Biosynthetic Reactions of Starch Synthesis

I. Introduction 33
II. Pioneering Studies 33
III. The ADPglucose Pathway Is the Major Pathway of Starch Synthesis *in Vivo* 34
IV. Alternative Pathways 37
V. Rate of Starch Synthesis versus Activities of the Starch Biosynthetic Enzymes 38
VI. A Missing Step? 40
VII. Summary 40

Synthesis of the Glucosyl Donor: ADPglucose Pyrophosphorylase

I. Regulatory Properties 43
II. Physiologic Relevance of the ADPGlc PPase Regulatory Properties 46
III. Subunit Structure 47
IV. Structure–Function Relationships 49
V. Function of the Higher Plant ADPGlc PPase Subunits . 50
VI. Identification of the Substrate Binding Sites 51
VII. Cloning of the ADPGlc PPase Genes and Comparison of Their Sequences 58
VIII. Hydrophobic Cluster Analysis 68
IX. Transcription 72
X. Genomic DNA 73

Starch Synthases

I. Introduction 75
II. Soluble Starch Synthases 75
III. Starch Synthases Bound to the Starch Granule 78
IV. Isolation of the Waxy Protein Structural Gene 81
V. Studies of *Chlamydomonas reinhardtii* Mutants 85
Further Readings 87

Branching Enzymes

I. Introduction . . . 89
II. Assay . . . 89
III. Purification of Branching Enzyme Multiforms . . . 92
IV. Mode of Action . . . 93
V. How Many Genes for Three Maize-Branching Enzymes? . . . 95
VI. Other Species . . . 98
VII. Relationship between Structure and Function . . . 101

Open Questions and Hypotheses in Starch Biosynthesis

I. Initiation of Starch Biosynthesis . . . 107
II. How Is the Starch Granule Formed? . . . 110
III. A Complete Pathway . . . 111

The Site of Starch Synthesis in Nonphotosynthetic Plant Tissues: The Amyloplast

I. Microscopy and Immunocytochemical Studies . . . 116
II. Cell Fractionation . . . 118
III. Transport of Carbon into Amyloplasts . . . 119

Regulation of the Starch Synthesis Pathway: Targets for Biotechnology

I. Introduction . . . 125
II. Genetic Engineering . . . 125
III. Vectors . . . 126
IV. Protoplast Isolation and Transformation . . . 127
V. Plant Regeneration . . . 128
VI. Tissue- and Organelle-Specific Expression . . . 128
VII. Antisense Technology . . . 129
VIII. Other Uses of Gene Technology . . . 130
IX. Transformation of Plants with an *Escherichia coli* Allosteric Mutant *glg* C Gene Increases Starch Content 131

X. Are Other Starch Biosynthetic Enzymes Rate Limiting? 134
XI. Other Physiologic Effects of Manipulation of Starch Synthesis 135
XII. Conclusions 136
Further Readings 137

Starch Accumulation in Photosynthetic Cells

I. Introduction 139
II. The Reductive Pentose Phosphate Pathway 140
III. The Chloroplast as a Transporting Organelle 143
IV. Control of Carbohydrate Metabolism 144
V. Regulation of the ADPGlc Pathway in the Chloroplast 145
VI. Starch Synthesis in Young Leaves 148
VII. Synthesis of Starch and Sucrose in C_4 Plants 148
VIII. The Regulation of Starch Synthesis in C_4 Plants 150
IX. Starch in CAM Plants 150
Further Readings 152

Starch Degradation

I. Plant Amylases and Phosphorylases 153
II. Debranching Enzymes 154
III. The Pathway of Starch Degradation in Plants 155
IV. Starch Degradative Enzymes Located Outside the Chloroplast: Possible Function 156
V. Digestion of Starch in Humans 157
VI. Mechanism of Action of Amylases and Phosphorylases 159
Further Readings 160

Industrial Applications of Starch

I. Industrial Applications of Starch 163
II. Manufacture and Properties of Starch 164
III. Physical Analysis of Starch and Derivatives in the Industrial Setting 167
IV. Chemical Modification of Starch 168

V. Conversion of Starch into Sweeteners 169
VI. Biodegradable Polymers 169
Further Readings 170

REFERENCES 171

INDEX 195

PREFACE

Research in starch biosynthesis is likely to have a great impact on agriculture and industry in coming years. Although the original purpose of research into starch synthesis was not industrial application, it is an example of how science, while trying to answer fundamental questions, may lead to the manipulation of nature for beneficial purposes.

Although the basic studies of starch synthesis were carried out in England during the 1940s, and led to the discovery of phosphorylase and Q-enzyme (branching enzyme), the basis of our modern ideas originated in Argentina from the work of Luis F. Leloir and Carlos E. Cardini. They founded in 1947 the Institute for Biochemical Research and during the late 1950s established that nucleoside diphosphate glucoses were involved in the biosynthesis of both glycogen and starch. These pioneers, "refugees" from a university system decimated by a dictatorial government, achieved great scientific advancement under difficult and very modest conditions. They were supported by private citizens at a time when the government would only employ members of the ruling party. Leloir and Cardini's group discovered the starch synthase reaction, first with uridine diphosphate glucose (UDPGlc) as a glucose donor (de Fekete *et al.*, 1960, 1961) and then with adenine diphosphate glucose (ADPGlc, Recondo and Leloir, 1961). This group isolated ADPGlc from corn grains and discovered the enzyme ADPGlc pyrophosphorylase (Espada, 1962). For some recollections of those romantic but dangerous times, please see Paladini (1996).

Our aim in writing this book has been to provide an up-to-date account of the biochemistry and molecular biology of starch. The chemistry of the starch granule and the biochemistry, molecular biology, plant physiology, and genetics of plant starch synthesis are discussed, and the recent findings regarding the properties of the starch biosynthetic enzymes and the studies describing their localization in the plant cell are emphasized. The implications of these studies for the seed, biotechnology, and modified starch industries are also discussed. We concentrate mainly on developments published since 1992, discussed against an historical background. For many of the more important discoveries, the authors' names and the dates are included so that the reader is introduced to most of the important workers in the field. For the subjects treated more succinctly, such as starch structure and degradation, reviews and books are cited as further reading. At the end of the book we include numerous references to the original literature but have not tried to be comprehensive.

Most starch is used as food, but about one-third of the total production is employed in a variety of industrial purposes that take advantage of its unique properties. We include a chapter in which the commercial uses of starch and its chemical and physical processing are summarily discussed. Clearly, how the raw material is used is important for the scientist who works in the basic sciences. Much can be gained by increasing the starch content in some plants and/or by manipulating its quality (e.g., by modifying the ratio of amylose to amylopectin). Starch content has already been increased in tomato fruit and potato tubers by using recombinant DNA and molecular biology techniques, and in the not too distant future it should be possible to alter its composition.

This book has been written with a broad readership in mind: starch has always been an important product, but now the capacity to modify its structure and increase the starch content of crops is attracting the attention of the seed companies, the chemical industry, and the research agencies. Because global warming is likely to affect the starch content in some plant species—a change that would, in turn, affect photosynthesis—this subject is of interest to physiologists, ecologists, and environmental agencies. All of this new attention has increased the flow of research papers in the field. In the next few years many of the basic questions posed here will be answered, leading, we hope, to advances in biotechnology and benefits for all.

ACKNOWLEDGMENTS

We thank our colleagues Alberto Iglesias, Brian Smith-White, Hanping Guan, Miguel Ballícora, Y. Y. Charng, and the many others who contributed to the development of the concepts presented in this book. We also thank Michigan State University and the State of Michigan for their support of our research. MNS will always remember with gratitude Juana Tandecarz and Carlos Cardini, mentors in science and in life, who were sadly lost too early.

OCCURRENCE OF STARCH

I. INTRODUCTION

Starch is a plant reserve polysaccharide, an end product of carbon fixation by photosynthesis, in which D-glucose residues are linked predominantly by α-(1,4) glucosidic bonds. It is present in most green plants and in practically every type of tissue: leaves, fruit, pollen grains, roots, shoots, and stems. Starch has a negligible osmotic pressure and thus allows plants to store large reserves of D-glucose without disturbing the water relations in the cell. All fruits contain starch, but in many of them only traces can be detected, and in most of them the starch is restricted to the chlorophyllous layers. Bananas and plantains have a relatively high starch content, especially before the onset of the climacteric, when nearly 90% of the dry weight of the fruit is starch. Starch present in pollen grains provides the energy required during germination and tube growth.

II. SEEDS

Members of the Gramineae (grasses) produce dry, one-seeded fruits, called caryopsis, commonly referred to as kernels or grains. The caryopsis (Fig. 1) consists of a fruit coat or pericarp, which surrounds the seed and adheres tightly to the seed coat. The seed consists of an embryo (or germ) and an endosperm enclosed by a nucellar epidermis and a seed coat. The main site of starch synthesis and accumulation is the endosperm, whose cells are packed with starch granules that form within the amyloplasts. Some starch is deposited in the embryo and pericarp early in development but later disappears. The starchy endosperm provides carbon skeletons and energy to the germinating embryo. Starch normally accounts for 65%–75% of the dry weight of the caryopsis in the mature, dry state. The embryo and the pericarp contain little starch, and values for the endosperm alone exceed 80%. The contents and cell walls of the endosperm make up the flour after the drying and processing of the grains. The baking properties of the flour are determined not only by the starch but also by the cell proteins that constitute the gluten.

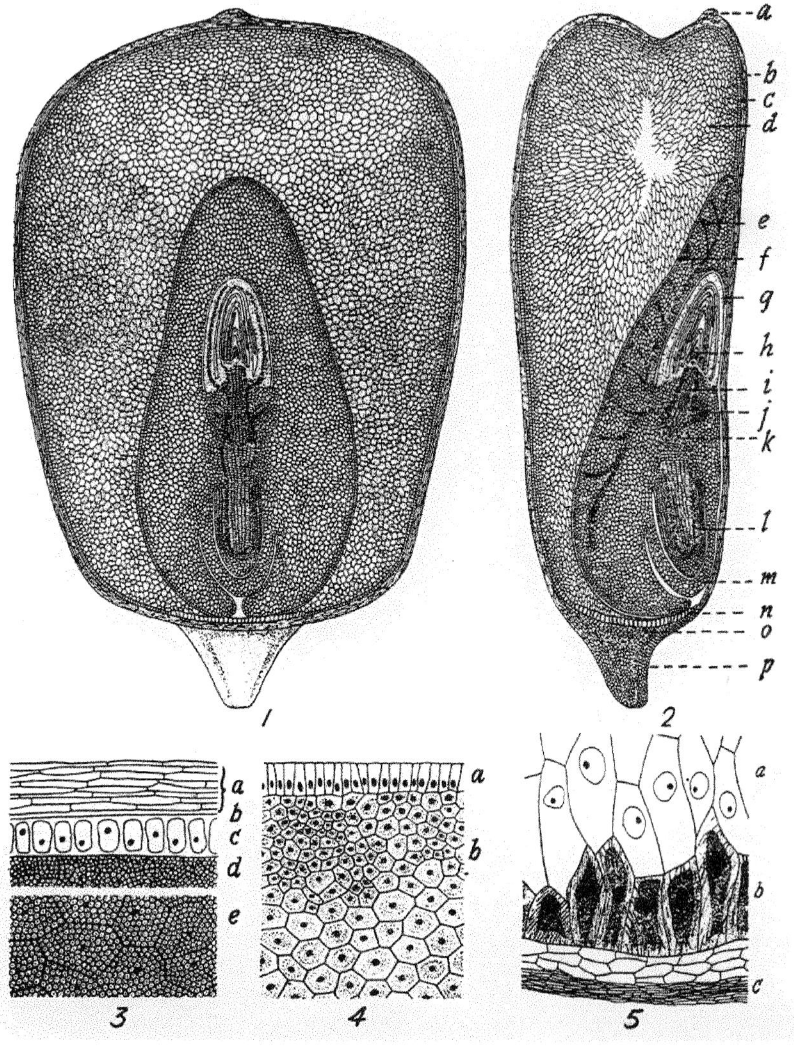

FIG. 1. The mature maize kernel. *1* and *2*, vertical sections in two planes of a mature kernel of dent corn, showing the arrangement of organs and tissues (magnification 7×); (*a*) silk (style) scar, (*b*) pericarp, (*c*) aleurone, (*d*) endosperm, (*e*) scutellum, (*f*) glandular layer of scutellum, (*g*) coleoptile, (*h*) plumule with stem and leaves, (*i*) first internode, (*j*) lateral seminal root, (*k*) scutellar node, (*l*) primary root, (*m*) coleorhiza, (*n*) lar node, (*o*) brown absission layer, (*p*) pedicel. *3*. Enlarged section through pericarp and endosperm (magnification 70×); (*a*) pericarp, (*b*) nucellar membrane, (*c*) aleurone, (*d*) marginal cell of endosperm, (*e*) inner endosperm cells. *4*. Enlarged section of scutellum (magnification 70×); (*a*) glandular layer, (*b*) inner cells. *5*. Vertical section of the basal region of endosperm (magnification 350×); (*a*) ordinary endosperm cell, (*b*) thick-walled conducting cells, (*c*) abcission layer. Figure reprinted with permission from Kiesselbach (1949).

The seeds of legumes have a lower percentage of starch than grass seeds: around 30% of dry weight for garden peas and 50% for cow peas. The study of the variations in seed morphology in maize and in peas, starting with Mendel, resulted in major contributions to the understanding of plant genetics. Some of these variations are caused by mutations affecting enzymes involved in the synthesis of starch and are discussed in the chapters corresponding to each enzyme.

III. STORAGE ROOTS AND TUBERS

Starch content in potato (*Solanum tuberosum*) tuber, in cocoyam corm (*Xanthosoma sagittifolium* and *Colocasia esculenta*), and in the roots of yam (*Dioscorea esculenta*), cassava (*Manihot esculenta*), and sweet potato (*Ipomea batatas*) ranges between 65 and 90% of the total dry matter, a result of a period of starch deposition that varies between 8 and 30 weeks. The dividing cells in newly initiated potato tubers, which are derived from stolons, contain little starch; however, once tuberization progresses, starch accumulation also progresses. Early in the development of the potato tubers, starch is distributed rather uniformly throughout the parenchyma. Later, two gradients of starch deposition appear and, as a result, the cortical parenchyma is richer in starch than the central part of the tuber, and the more mature, basal end of the tuber contains more starch than the younger distal tissues. Yams and cassava also display specific patterns of starch accumulation that are related to the particular pattern of differentiation of the organ.

IV. STARCH IN THE GRAVITATIONAL RESPONSE OF ROOTS AND STEMS

Sedimentation of amyloplasts within the cell has been correlated with the capacity of the plant to perceive gravity. The buoyant mass of amyloplasts present in specialized cells in the center of the root cap and in the stem (depending on the plant species, in the endodermis, the bundle sheath, or in the parenchyma to the inside of the vascular bundle) would allow the amyloplasts to sediment inside the cell, where the cytosol would have a relatively low viscosity. This sedimentation would translate into a signal of an unknown nature, maybe through pressure onto a sensitive part of the cell or acting as a mechano transducer, etc. Whatever the nature of the signal, it eventually results in the asymmetry of the organ and its curvature. The isolation of starchless mutants of *Arabidopsis thaliana* and *Nicotiana sylvestris* has made it possible to compare the gravitational responses of plants differing only in the amount of starch, as plastids are present in both wild-type and starchless mutants. Although it was initially believed that the responses were identical (Caspar and Pickard, 1989), apparently the starchless mutants in both species are less sensitive to gravity (Sack and Kiss, 1989).

V. LEAVES

In leaves, starch is deposited in granules in the chloroplasts during active carbon dioxide fixation by photosynthesis throughout the day and is degraded by respiration at night. Starch remobilization ensures the constant availability of photosynthates to the whole plant. Mutants of *A. thaliana* that are able to synthesize sucrose but unable to synthesize starch grow at the same rate as the wild type in a continuous light regime, but growth rate is drastically reduced if they are grown in a day–night regime (Caspar *et al.,* 1986). The biosynthesis and degradation of starch in the leaf are, therefore, more dynamic than the metabolism in reserve tissues. Chloroplast starch granules are smaller than those in reserve tissues and their shapes are not species specific and are likely to be determined simply by the space available at the site where they are formed.

VI. GREEN ALGAE

The presence of starch has been demonstrated in several species of green algae (*Chlorophyceae*). Starch content in four genera of green algae studied by Love *et al.* (1963) contained about 1% starch. The viscosity of algal starch solutions was lower than that of potato starch, indicating a lower degree of polymerization, but the percentage of amylose was not very different. Extraction of algal starch is complicated by the presence of a large amount of other polysaccharides, especially sulfated ones.

Algae lack differentiated organs and one would expect the role of starch and its structure to resemble those of leaf starch rather than those of reserve tissues. In this decade, a green algae, *Chlamydomonas reinhardtii,* has become a system of choice for the study of starch synthesis. Ball and his collaborators (1990) studied this algae under sets of conditions that favor accumulation of "storage" starch (N depletion, dark, carbon, and energy supplied as acetate) or "photosynthetic" starch (light, complete nutrient solution). The structure and site of accumulation within the cells vary according to the growth conditions.

VII. OTHER RESERVE POLYSACCHARIDES

Starch is not the only storage polysaccharide found in plants. A storage substance is one that can be broken down rapidly to provide energy and/or building blocks for new growth by respiration. Reserve polysaccharides are stored in plastids (as in the case of starch), in the cell vacuole, or outside the plasmalemma, in the cell-wall region. The presence in the plant of enzymes capable of degrading the substance is a good indicator of its role as reserve. This definition can be applied with ease to starch in higher plants or to glycogen in cyanobacteria, but for other polysaccharides found in some algae, the role is less clear (Percival and McDowell, 1985). For example, xylans—polymers of xylose present in Rhodophyta, the red algae, and in Chlorophyta, the green algae—may fulfill more than a single function in the same algae (i.e., as reserve and as part of the cell-wall structure). Cell-wall polysaccharides in some senescent tissues, such as ripening fruits, can be turned over and the monosaccharides produced can be incorporated into polysaccharides. An arabinogalactan mucilage present within the style canal of *Lilium* acts as a source of carbohydrate precursor for the growing pollen cell wall (Loewus and Labarca, 1973).

Laminarin, a linear glucan containing mainly β-D-(1→3) linked glucose, with some β-D-(1→6) branching points, is found in *Laminaria,* a brown seaweed.

Mannans, in which mannose units are linked predominantly in β-D-(1→4) bonds, are found in the red seaweed *Porphyra umbilicalis,* in the seed of the tagua palm (*Phytelephas macrocarpa*) in the form of massive thickening of the cell walls of the endosperm, and in the endosperm of members of the Umbelliferae and of the Compositae (e.g., lettuce seed).

Other reserve glucans have been described (Meier and Reid, 1982), but in higher plants only starch and fructan, a water-soluble polymer of D-fructose that is osmotically active, are widespread. Hendry (1987) estimated that fructans are present in about 12% of vascular plants, many of them from temperate climates. It has been proposed that fructans, which are located in the cell vacuole and are osmotically active, can decrease the freezing point of the cell sap, slow the rate of freeze-dehydration, and afford frost hardiness to the plants that store them. Long-term storage of fructan can occur in specialized organs (e.g., the tubers of the Jerusalem artichoke) (Jefford and Edelman, 1961), in the stems and developing inflorescences of temperate grasses and cereals during periods of reproductive development (Archbold, 1940), and in the seeds of some Gramineae during the early stages of grain development, before starch synthesis begins. Pollock and Chatterton (1988) discussed the possible advantages afforded to plants by fructan accumulation in leaves as compared to starch.

Floridean starch containing α-D(1→4), α-D-(1→6), and possibly some α-D-(1→3) bonds is the characteristic reserve polysaccharide in the Rhodophyceae (red algae) and is present as granules in the cytosol. The presence of α-D-(1→3) bonds, if confirmed, would clearly differentiate floridean starch from both starch and glycogen, but they could be an artifact. Floridean starch has been detected in many species of red algae (Meeuse *et al.*, 1960) but has been characterized in only a few cases. In its viscosity and molecular weight (MW) of approximately 10^8, it resembles amylopectin (Greenwood and Thomson, 1961), but in other respects, (e.g., average chain length) it resembles glycogen (although chain lengths can vary from about 10 to 18).

Glycogen, an α-1,4-glucan with α-1,6 branching points, is the storage polysaccharide for cyanobacteria (blue-green algae). Cyanobacteria are prokaryotes and, although they are photosynthetic, they have no plastids and their glycogen is present as small granules in the cytosol. In thin sections seen under the electron microscope, they appear as spheres of 25 to 30-nm-diameter or rods (31 by 65 nm in *Nostoc*) that stain densely with lead citrate and are often located between the thylakoids and are more prominent in nitrogen-limited photosynthesizing cells (Shively, 1988).

VIII. EXPERIMENTAL SYSTEMS IN THE STUDY OF STARCH METABOLISM

The model experimental systems mentioned more frequently in this book are the kernels of maize and rice, the potato tuber, the pea seed, the aerial parts (leaves and stem) of *Arabidopsis thaliana,* and the alga *Chlamydomonas reinhardtii.* Some of these systems (e.g., rice, potato) have been chosen by researchers for their economic importance, whereas other plants have been chosen because many mutant lines are available for study (e.g., pea) or because they are particularly amenable to genetic studies (*Arabidopsis*). It should not be expected, however, that these few species represent "perfect" models (if such a thing exists) of how starch synthesis operates in plants in general, and one should be cautious when extrapolating to other species the information obtained using one system. For example, potato and maize have been selected for centuries in the search of high starch production, and we could expect that breeding has introduced some peculiar characteristics leading to high starch accumulation that may not be typical of what the species was before domestication. However, *Arabidopsis* is a good system in the sense that it has not been subject to selective pressure, but the plant is very small, making biochemical studies difficult and limited mostly to the leaves.

It is worth noting that bread wheat (*Triticum aestivum*), one of the most important world crops, is far from an ideal experimental system. Wheat is a natural allopolyploid. It has 21 pairs of chromosomes, which represent three sets of chromosomes that come from three different wild relatives, possibly *T. monococcum, T. searsii,* and *T. tauschii.* The bread wheat as we know it is the result of a combination of naturally arising mutations, such as the gene *Ph* that allows the coexistence of the three related sets of chromosomes, and cultivation by humans for more than 10,000 years. Breeding has resulted in a very high harvest index; that is, a gradual increase in the proportion of above-ground assimilates going to the grains, the harvested sink organs. The molecular bases for this ever-increasing harvest index are probably related to increased starch synthesis selected by breeding. However, the hexaploidy of wheat makes genetic manipulation complicated, and biochemical study of the kernel enzymes is also difficult.

A. MAIZE

Maize (*Zea mays*) is a cross-pollinated plant that has evolved (with great help from humans) into thousands of varieties or races that are composed of a great deal of genetic variability; the wild relatives of maize are teosinte (*Zea mexicana*) and *Tripsacum.* The maize cultivated in commercial agriculture represents a very small fraction of this genetic variability and consist of a few hybrids obtained by the systematic crossing of a few inbred lines. Besides its commercial importance, another reason why maize is frequently used as a model system is that it bears male and female flowers on separate structures (Fig. 2). This characteristic facilitates controlled pollinations and genetic studies, and also the outcrossing responsible in part for the enormous genetic variability of the species. Maize produces a large ear with 500 or more individual kernels (the main site of starch deposition), each containing a prominent endosperm and a large embryo, facilitating biochemical studies. There is also a large amount of data available on the physiology of the whole plant and its ultrastructure, and maize is the most extensively characterized flowering plant from a genetic and cytogenetic point of view.

The development of the kernel following fertilization takes 40–50 days and is accompanied by a 1400-fold increase in the volume of the embryo sac; the growth of the embryo and accumulation of food reserves in the endosperm is completed by about day 40. A mature kernel has three parts: pericarp, endosperm, and embryo (Fig. 1). The pericarp, the tough, transparent, outer layer of the kernel, is derived from the ovary wall and is, therefore, genetically identical to the maternal parent; the endosperm and embryo represent the next generation.

FIG. 2. The maize plant. (Classic drawing by W. C. Galinat.)

Besides the usual forms of genetic change present in other plants (i.e., gene mutation and recombination), transposable genetic elements, also called *jumping genes,* are an additional source of genetic variation in maize. These are genetic elements that can occasionally move (transpose) from one position in the chromosome to another position in the same chromosome or in a different chromosome. Transposable elements can mediate chromosomal rearrangements, and were first discovered in maize by M. Rhoades, where they manifested themselves as unstable mutant alleles, i.e. alleles for which reverse mutation occurs at a very high rate. In the 1950s, Barbara McClintock found a genetic factor *Ds* (*Dissociation*) that causes a high tendency towards chromosome breakage at the location in which it appears. Controlling elements in maize can inactivate the gene in which they reside, cause chromosome breaks, and transpose to other locations within the genome. Complete elements can perform these functions unaided; other forms with partial deletions can only transpose with the aid of a complete element located elsewhere in the genome.

One locus related to starch synthesis, *waxy,* has been the object of intense study on the effects of the *Ds* element. The *Ds* element can move into a gene making it into an unstable mutant dependent on the other element, Ac. The *wx* locus is one example and was studied in detail by Oliver Nelson, who paired many different unstable wx alleles in the absence of the *Ac* mutation. He then screened the heterozygotes for the rare wild-type recombinants by staining the pollen with iodine reagent (*Wx* pollen, containing normal starch, stains black; *wx* pollen, lacking amylose, stains red) and, by counting the frequency of the wild-type recombinants, he obtained a fine structure map of the waxy gene. Nelson also showed that the different mutable waxy mutant alleles were caused by the insertion of the *Ds* element in different positions within the *waxy* gene.

Maize is a particularly favorable material for the investigation of the biochemical effects of genetic lesions because of the large size of its seeds and because of the translucent pericarp, which allows detection of any deviation from normal development. The substantial background of genetic information is also very helpful. Some of the mutants available for study are *amylose extender, dull, sugary 2,* and *waxy,* all of which affect the ratio of amylopectin to amylose. The *shrunken-1, shrunken-2,* and *brittle-2* mutations reduce starch content of the endosperm. The *sugary-1* mutant is unique in that the principal storage polysaccharide is not starch but the highly branched and water soluble phytoglycogen. Besides the mutants that have been biochemically characterized, O. Nelson (1985) mentions many more mutants (not allelic to those mentioned previously) that even now are awaiting identification.

B. POTATO

The potato plant (*Solanum tuberosum*) is bushy, sprawling, and dark green, with compound leaves that resemble those of a close species, the tomato. The leaves are arranged in a spiral around the stem, and the flowers are arranged in clusters. They are about 1-inch wide and 5-petaled, and range in color from white to pale blue to purple. The plant is completely poisonous except for the tubers; indeed, all plant members of the nightshade family, which includes potatoes, tomatoes, and eggplants, contain the poisonous alkaloid called solanine, a natural defense against its many predators.

The life cycle of the potato plants cultivated today is completely asexual (i.e., tuber to sprout to plant to tuber). When rapid leaf growth slows down, the plant begins to form flowers, and underground stems (stolons) begin to branch out and swell at their tips. Sucrose is sent from the mature leaves, the sources, to the rest of the plant and the stolons, the sinks. The starch is deposited at the ends of the stolons, forming tubers.

C. ARABIDOPSIS THALIANA

The cruciferous weed *Arabidopsis thaliana* has become a model system for the study of an unusually wide variety of aspects of plant biology. *Arabidopsis thaliana* is a small weed, related to the mustards, and possesses a number of characteristics that make it an ideal object of genetic study. It has a rapid life cycle, passing from germination to flowering and setting of seeds in about 5 weeks; the plant may be self- or cross-pollinated, facilitating genetic analysis. The small size of the plants facilitates its cultivation of large numbers in laboratory conditions and the screening for relevant mutants after chemical mutagenesis. Another advantage is that it is relatively easy to transform some lines of *Arabidopsis thaliana* using the *Agrobacterium* Ti plasmid. The *Arabidopsis* genome is relatively small, with about 10^8 bp of DNA, and most of this genetic material is single copy sequences, facilitating the development of a very detailed genetic map.

D. ANTIRRHYNUM MAJUS

Antirrhynum majus is a common cultured garden plant, the snapdragon. The normal typus or wild type of *A. majus* is defined to be the Sippe 50 strain.

Gene inactivations and reactivations caused by the insertion and excision of transposable elements of the kind first discovered in maize, also appear in *Antirrhynum,* facilitating the identification and molecular analysis of genes involved in flower development and organ identity. Although in *Antirrhynum* the best studied genes are those involved in the synthesis of pigments and in flowering, it is now being used in the investigation of the mechanism of starch biosynthesis by Romero and colleagues.

Gene disruption is an experimental tool used for "reverse genetics," in which a gene is specifically inactivated, as pioneered in yeast, so that the precise function of the gene may be determined. A "cryptic" DNA or protein sequence is used to discover the normal role of the gene at the phenotypic level. Another gene with a selectable function can be inserted into the middle of a wild-type allele of the gene of interest carried on a plasmid. A linear derivative of such a construct would insert itself specifically at the wild-type locus, automatically disrupting it and at the same time allowing the selection of the recombinant via the selectable gene. In the case of starch biosynthesis, study is still limited to the specific effects of the relevant genes for which mutants have been obtained, but the use of gene disruption in plants such as *Antirrhynum* would greatly expand the options available to the biochemist in search of the role of enzymes of starch metabolism and multiforms in the final architecture of the starch granule.

E. CHLAMYDOMONAS REINHARDTII

Although cyanobacteria (also called *blue–green algae*) are often used as a model system for plants because they are photosynthetic, they are prokaryotic and more similar to bacteria than to plants in many ways. Cyanobacteria, for example, accumulate glycogen rather than starch and have no organelles. Conversely, *Chlamydomonas reinhardtii,* a unicellular organism used since 1990 in the study of starch synthesis is a green algae, is a better system to study the effect of mutations in the relevant enzymes on starch structure. *Chlamydomonas* is a large genus of green flagellates; more than 600 species have been described worldwide from marine and freshwaters, soil, and even snow. Until the 1970s, *Chlamydomonas* was considered by many to be the most ancient of the green plants, but according to the current opinion they are considered nonancestral members of the chlorophyte lineage (Chlorophyceae) of green algae. Several species of *Chlamydomonas* have become important experimental organisms in fields such as cell and molecular biology, genetics, plant physiology, and biotechnology.

Swimming cells have a single nucleus and two flagella inserted into a minute papilla at the anterior end of the cell; the cell wall is thin. Most of the cell volume is occupied by one or more grass-green chloroplasts. In the most frequently used species, *C. reinhardtii,* only one cup-shaped chloroplast is present; one or more pyrenoids are present within the chloroplast; starch grains surround the pyrenoid.

Vegetative cells are usually haploid, and reproduce asexually by division into two, or some small multiple of two, progeny cells. Under certain conditions, usually involving induction of vegetative cell growth under nitrogen limitation, vegetative cells divide to form gametes. Gametes look like vegetative cells, but have differentiated mating structures near their apices. Cysts are usually diploid, formed by fusion of gametes. Meiosis in the cysts usually yields four vegetative cells. The life cycle of *Chlamydomonas* is easy to manipulate under controlled culture conditions.

FURTHER READINGS

These sources provide additional in-depth coverage of this topic. For complete reference, please see the Reference section at the end of the book.

Brinson, K., and Dey, P. M. (1985)
Jenner, C. F. (1982)
Manners, D. J., and Sturgeon, R. J. (1982)
Meier, H., and Reid, J. S. G. (1982)
Neuffer, M. G., Coe, E. H., and Wessler, S. R. (1997)
Percival, E., and McDowell, R. H. (1985)
Pollock, C. J., and Chatterton, N. J. (1988)
Pontis, H. G., and del Campillo, E. (1985)
Sack, F. D., and Kiss, J. Z. (1989)
Sheridan, W. (1982)
Shively, J. M. (1988)

PHYSICOCHEMICAL STRUCTURE OF THE STARCH GRANULE

I. THE STARCH GRANULE

Starch and glycogen (the storage material in animals and bacteria) are both polymers of α-D-glucose, but starch differs from glycogen in that starch consists of a highly ordered and dense packing of glucan chains organized within large, insoluble granules. The starch granules are formed in the amyloplast (see the chapter, "The Site of Starch Synthesis in Nonphotosynthetic Plant Tissues; The Amyloplast"), specialized in the synthesis and long-term storage of starch, or in the chloroplast (see the chapter, "Starch Accumulation in Photosynthesis Cells"), where starch serves as a temporary store of energy and carbon.

Starch granules vary in size, shape, composition, and properties (Table I), and they are a semicrystalline material. Because the starch granule has a high degree of order, when viewed in polarized light it shows birefringence, the maltese cross of Fig. 1.

The shape and size of the granules depend on the source. For example, pollen starch granules are about 2 μm in diameter and those from canna starch have diameters of up to 175 μm. Although the microscopic appearance of starch granules (Fig. 2) is sufficiently characteristic to allow the identification of the botanical source of the polysaccharide, in each tissue there is a range of sizes and shapes. For example, in barley starch there are two populations of granules: one is composed of large lenticular granules with diameters between 15 and 35 μm, and another of small spherical granules with diameters between 1 and 10 μm. In general, the diameter of the starch granule changes during the development of the reserve tissue. In addition to size and shape, there are also some fine features that are characteristic of each species (e.g., the "growth rings" seen in potato starch), which help to identify the botanical source of the starch upon microscopic examination.

II. AMYLOSE AND AMYLOPECTIN

At least two polymers can be distinguished within the starch granule: amylose, which is essentially linear; and amylopectin, which is highly

TABLE I

COMPARISON OF STARCHES USED COMMERCIALLY[a]

Type of starch, composition and properties	Potato	Maize: Wild type	Maize: Waxy	Wheat	Tapioca
Starch granules					
Shape	Oval–spherical	Round–polygonal	Round–polygonal	Round–lenticular	Round–polygonal
Diameter, range (μm)	5–100	2–30	2–30	0.5–45	4–35
Composition					
Moisture[b]	19	13	13	13	13
Lipids[c]	0.1	0.8	0.2	0.9	0.1
Nitrogen compounds[c]	0.1	0.35	0.25	0.4	0.1
Ash[c]	0.35		0.1	0.2	0.1
Phosphorus[c]	0.08	0.02	0.01	0.06	0.01
Starch-bound phosphorus[c]	0.08		0	0	0
Pregelatinized starches					
Taste and odor substances	Low	High	Medium	High	Very low
Amylose					
Amylose content[c]	21	28	1	28	17
Degree of polymerization (DP)					
Number average DP	4900	930	—	1300	2600
Weight average	6400	2400	—	—	6700
Apparent DP distribution	840–22,000	400–15,000	—	250–13,000	580–22,000

Amylopectin					
Degree of polymerization (DP)					
DP × 10^4 (range)	0.3–3	0.3–3	0.3–3	0.3–3	0.3–3
Gelatinization					
Pasting temperature, C°	60–65	75–80	65–70	80–85	60–65
Swelling power at 95°C	1153	24	64	21	71
Solubility at 95°C	82	25	41	48	23
Starch pastes					
Paste viscosity	Very high	Medium	High	Low	High
Water binding[d]	24	15	22	13	20
Paste texture	Long	Short	Long	Short	Long
Paste clarity	Nearly clear	Opaque	Fairly clear	Cloudy	Quite clear
Resistance to shear	Low	Medium	Low	Medium	Low
Rate of retrogradation	Medium	High	Very low	Medium	Low
Main commercial uses	Food, paper adhesives	Sugar, paper, corrugated board	Food, adhesives	Sugar, bakery	Food, adhesives

[a] Data from Swinkels (1989).
[b] Moisture at 65% RH and 20°C.
[c] % of dry matter.
[d] Water-binding capacity in parts of water per part of dry native starch to reach similar hot viscosity after cooking.

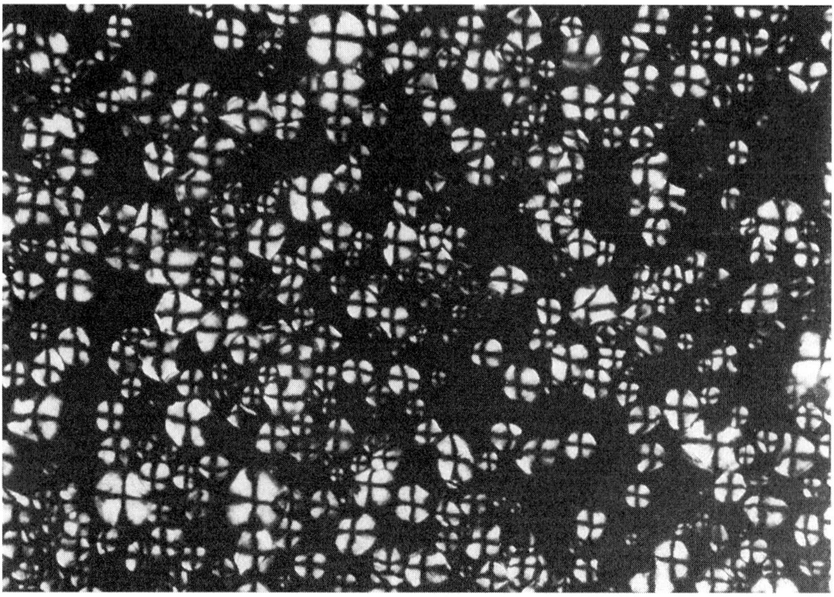

FIG. 1. The birefringence (maltese cross) shown by maize starch illuminated with polarized light, 700×. From Fitt and Snyder (1984).

branched (Fig. 3, Table II). Amylose is found mainly as linear chains of about 1500 units of α-D-glucopyranosyl residues linked by α-(1→4) bonds (molecular weight around 250,000; the molecular weight of an anhydroglucose residue is 162), but the number of anhydroglucose units varies widely with plant species and stage of development. Some molecules found in the amylose fraction are branched to a small extent (1→6 α-D-glucopyranose; 1 per 1000 or 1500 glucose residues). In contrast, amylopectin, which usually constitutes about 70% of the starch granule, is more highly branched, with about 4 to 5% of the glucosidic linkages being α-1→6 (Fig. 3). Methylation followed by acid hydrolysis shows that there is one nonreducing end group for every 20 to 25 D-glucose residues; this has been confirmed by the periodate oxidation method. These results are only compatible with a highly branched molecule and explain why amylopectin does not form threads or films in the same way as amylose. From the hydrolysis products, about 3% are 2,3-di-*O*-methyl-D-glucose, indicating that some glucose residues are joined to others through $C_{(6)}$ as well as through $C_{(1)}$ and $C_{(4)}$, and these units constitute the branch points. This is confirmed by the isolation of isomaltose and panose (α-D-G_p1–6α-D-G_p1-4-D-G_p) after partial hydrolysis of amylopectin. Thus, the average chain length of amylose is about 1500

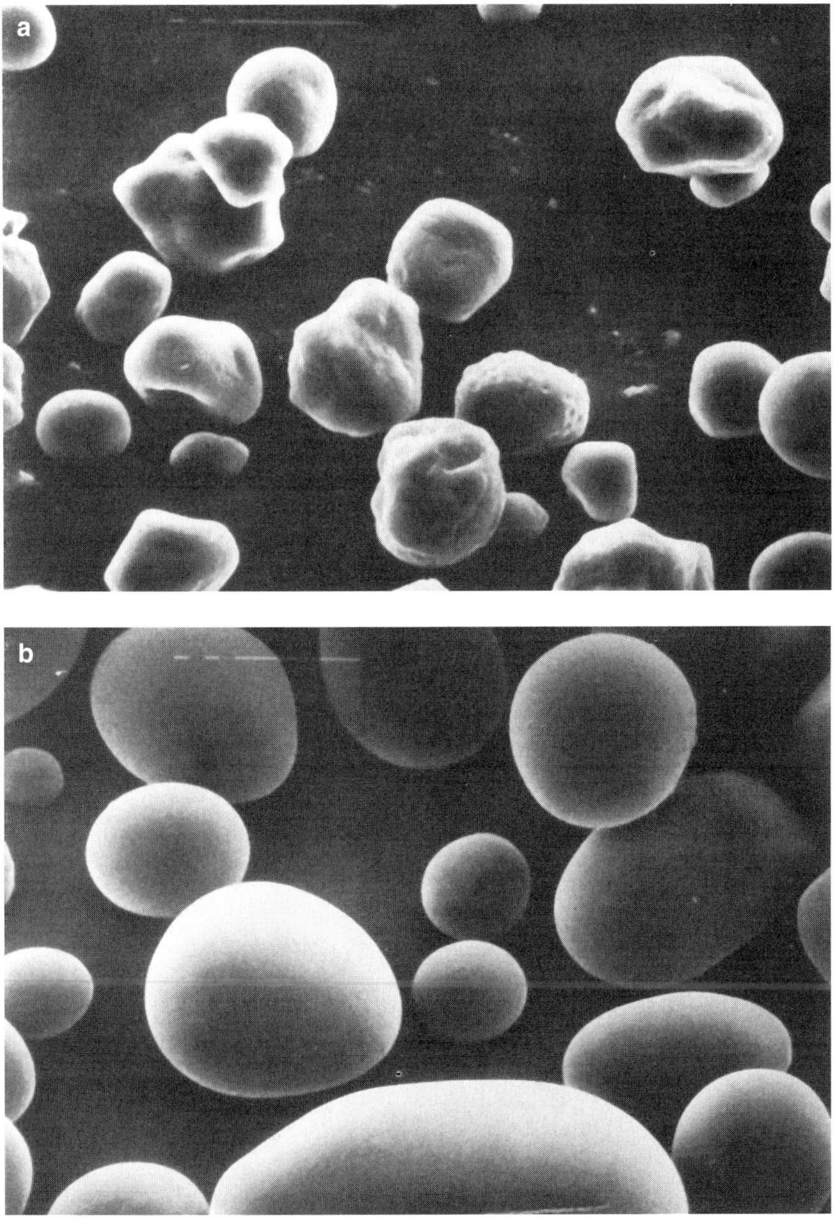

FIG. 2. Scanning electron micrographs of starch granules from (a) maize, 1500×; (b) potato, 1500×; (c) rice, 5000×; and (d) tapioca, 1500×. From Fitt and Snyder (1984).

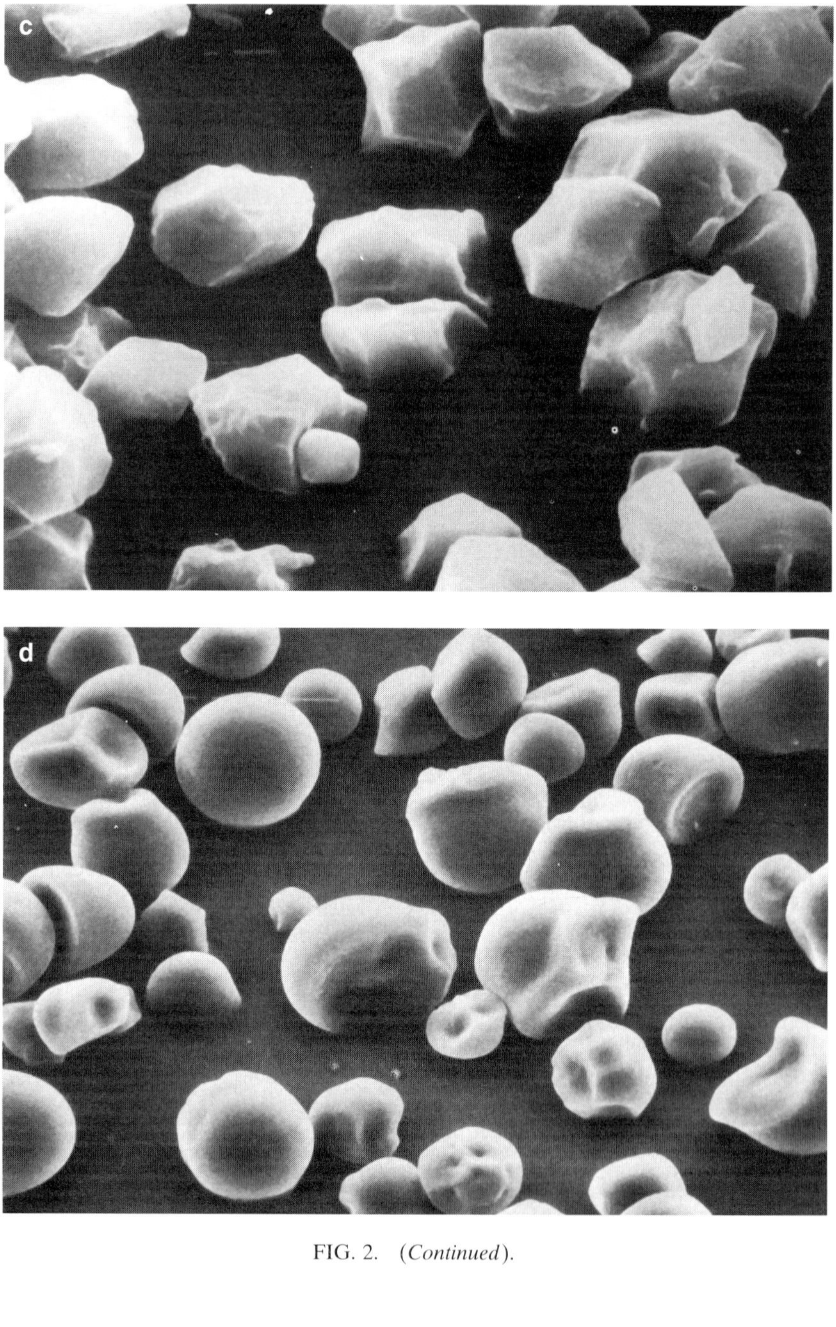

FIG. 2. (*Continued*).

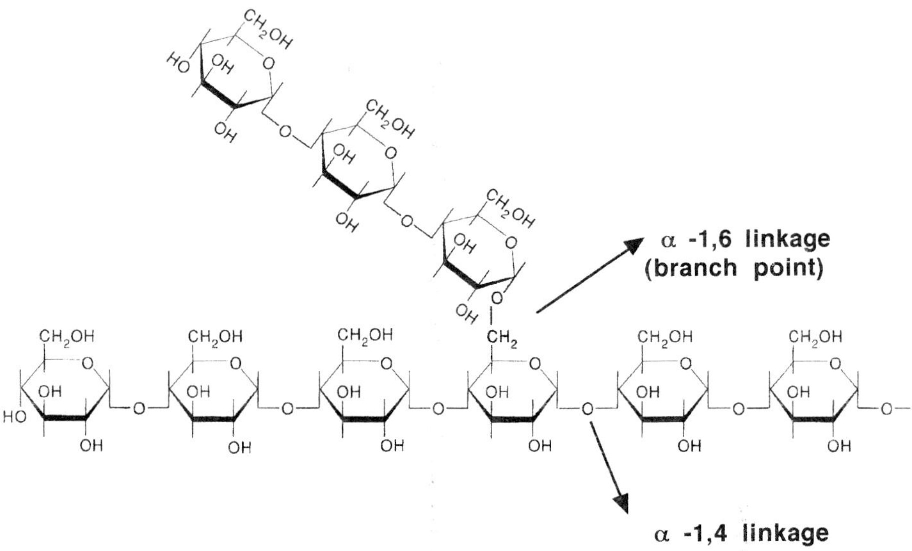

FIG. 3. The α-(1,4) and α-(1,6) glycosidic linkages between the glucosyl units present in starch.

TABLE II

PROPERTIES OF THE STARCH COMPONENTS AMYLOSE AND AMYLOPECTIN[a]

Property	Amylose Whole	Amylose Linear	Amylose Branched	Amylopectin	Intermediate fraction
Branch linkage (%)	0.2–0.7	0	0.2–1.2	4.0–5.5	2–3.5
Average chain length (CL)	100–550	800	140–250	18–25	30–50
Average degree of polymerization (DP)	700–5000		10^3–10^4	10^4–10^5	10^2–10^4
λ_{max} (nm)	640–660			530–570	570–580
Blue value[b]	1.2–1.6			0–0.2	0.3–0.7
Iodine affinity (g per 100 mg)	19–20.5			0–1.2	2–10
Helix formation[c]	Yes	Yes	Yes	No	No
β-Amylolysis limit	70–95	100	40	55–60	57–75

[a] Data from Hizukuri (1995).
[b] Blue value: absorbance at 680 nm of the iodine complex in controlled conditions.
[c] With 1-butanol.

glucose residues and, for amylopectin, the average chain length is about 20 to 25 units. A typical molecular weight for amylopectin is around 10^8, with about 600,000 glucose residues.

It should be noted that the different structures of amylose and amylopectin confer distinctive properties to these polysaccharides (Table II). The linear nature of amylose is responsible for its ability to form complexes with fatty acids, low-molecular-weight alcohols, and iodine; these complexes are called clathrates or helical inclusion compounds. This property is the basis for the separation of amylose from amylopectin: when starch is solubilized with alkali or with dimethylsulfoxide, amylose can be precipitated by adding 1-butanol and amylopectin remains in solution.

When an aqueous starch solution is left to stand for some time, partial precipitation occurs. This is known as retrogradation and is due to the separation of the amylose fraction. The linear molecules align themselves parallel to each other and become held together by hydrogen bonds. The aggregates increase in size until they exceed colloidal dimensions and therefore precipitate. Because of this tendency, it is difficult to work with amylose, and to keep it in solution, it is often necessary to keep it at a high pH and at relatively high temperatures. Conversely, amylopectin does not generally form complexes and is stable in aqueous solutions. In some plant varieties, a minor third fraction, referred to as "anomalous amylopectin" or "intermediate fraction" (Table II), may also be present and can complicate fractionation. This fraction has fewer branch linkages than normal amylopectin; that is, it has greater average chain length (Hizukuri, 1995).

The early work of Katz and colleagues in the 1930s established that starch can give a number of distinct types of X-ray patterns, depending on the source of the starch and the treatment to which the granules were subjected. In intact starch granules, three dominant patterns, named A, B, and C, can be observed (Fig. 4). In the 1940s, French and his co-workers, using flow dichroism and X-ray examination of the amylose–iodine complex, showed that the amylose molecule is in the form of a helix, as had been proposed earlier by Hanes. French *et al.* suggested that there were six D-glucose units in each turn, with the iodine atoms lying along the axis of the helix. In 1972, Kainuma and French pointed out that models based on a sixfold helix could not satisfy the experimental values obtained by X-ray crystallography for B-amylose, and they postulated the presence of double helices. In solutions containing suitable "guest" molecules, segments of amylose would complex to form single left-handed V-type helices with a hydrophobic cavity of about 0.5 nm in diameter. In I_2/KI solution, the guest molecules are polyiodide ions (mostly I_3- or I_5-). The color and λ_{max} of the complexes vary with chain length and analytic conditions, and the iodine binding capacity is around 20 g/100 g amylose.

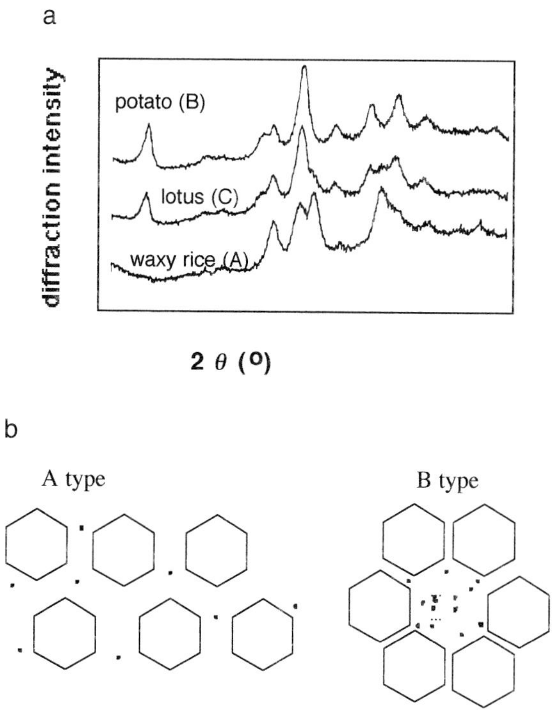

FIG. 4. (a) Diffractometer patterns of starch showing typical A, B, and C types of X-ray spectra. (b) Packing of double helices in the crystalline patterns proposed for the A and B types of starch. The C type would be a mixture (in varying amounts for different species) of A and B type of packing. After Hizukuri (1995).

The capacity of starch to stain blue-black with iodine suggests that some of the amylose is present in the starch in the V-form. The lipids present in cereal starch would bind to amylose if it were in the V-form, and yet X-ray analysis does not show the presence of the V-polymorph in cereal starches (i.e., most of the amylose would be in the amorphous form). The conclusion is that although a significant part of the amylose is probably in the helical form, the three-dimensional order necessary to give a crystalline diffraction pattern is absent. Indeed, the crystalline nature of starch is now attributed to the presence of amylopectin and not to amylose. Starch from waxy mutants contains only amylopectin (and no amylose), but this starch has the same degree of crystallinity and the same X-ray pattern as the regular starches that contain both components.

Starch granules are microcrystalline, comprising crystalline domains, non-crystalline domains, and possibly transitional regions. Native starch granules have crystallinities estimated to range between 20 and 40%; this relatively low crystallinity is responsible for the low-quality X-ray diffractograms. Although it is generally thought that branching in a molecule is detrimental to crystallization, it seems that in the case of starch, amylopectin, which is the branched molecule, and not the almost linear amylose, is the fraction responsible for the crystalline nature of starch. Indeed, Hizukuri (1985) found that the chain length of amylopectin is a basic factor in the determination of the crystalline type of the starch.

On the basis of the double helix concept (Kainuma and French, 1972), several molecular models have been proposed for the unit cell structures that would satisfy the X-ray and electron diffraction experimental data. As proposed by Imberty *et al.* (1987, 1988), the double helices in both A and B types would be identical, but the mode of packing of the helices and the water content would differ (Fig. 4b). The A and B patterns represent true crystalline forms of starch, but the C form is a composite, containing elements of A and B. Many different structures have been proposed to explain the crystalline patterns (Banks and Muir, 1980; French, 1984), but it seems that the patterns are a result of a combination of factors, including the chain length of the amylopectin, helix packing, and water of crystallization (Hizukuri, 1986). The A pattern is more frequent in cereal starches, whereas the B pattern is found in potato and amylomaize starch. The C pattern can be obtained by mixing maize and potato starches (Hizukuri *et al.*, 1961), but it is also found in nature—for example, in smooth-seeded peas and in bean starches. Heat-moisture treatment can change the X-ray diffraction pattern from the B to the A pattern. Plants producing starch giving a B pattern can produce starch with an A pattern if they are grown at higher temperatures or if the isolated starch is partly dehydrated. The crystallinity of starch granules can be destroyed mechanically; for example, ball milling at room temperature will destroy both the birefringence and the X-ray pattern.

The orientation of the principal axis of the crystallites is radial with respect to the hilum (center) of the granule (French, 1972). Small-angle X-ray scattering data suggest the existence of a 9-nm repetitive unit that is found in all plants, implying the presence of a highly ordered biosynthetic pathway that is well conserved throughout the plant kingdom (Jenkins *et al.*, 1993). This repetitive unit is composed of an amorphous and a crystalline lamella. Although the sum of both lamellae remains constant (9 nm), the relative size of each in the repetitive unit is under genetic control. Lengths of 4 to 6 nm have been reported for the size of the crystalline lamella, and this would amount to a linear α-1,4-glucan of a size ranging from 12 to 18 glucose residues. Powder diffraction patterns of native starch have been used to determine the three-dimensional structures of the crystalline lamella (reviewed by French, 1984; Imberty *et al.*, 1991; Hizukuri, 1995), and three types of diffraction patterns (A, B, C) were obtained. Each of these patterns is interpreted as the packing of linear (unbranched) parallel glucan double helices.

Amylopectin molecules are very large, flattened disks consisting of α-(1,4)-glucan chains joined by frequent α-(1,6)-branch points (Fig. 3). The chain that contains the single reducing end group is called the C-chain, to which all the other chains are ultimately attached (Fig. 5). The A-chains carry no branch points and are attached to B-chains, which have one or more branch points and are themselves attached to other B-chains or to the one C-chain (Peat *et al.*, 1952). Many models of amylopectin structure have been proposed (Fig. 5a), but of these the most satisfactory models, those that fit the experimental data available, are those proposed by Robin *et al.* (1974), Manners and Matheson (1981), and Hizukuri (1986; Fig. 5b). The arrowheads indicate the presence of a branching point [i.e., an α-(1,6) bond], and the branched regions of amylopectin are amorphous. The potentially crystalline clusters of A- and B-chains—the short, linear chains beyond the branch points that can form left-handed, parallel-stranded double helices—are also shown. The size of the crystallites is derived from the average chain length determined experimentally, and the ratio of A- to B-chains in the model can also be measured by enzymatic hydrolysis. Highly purified forms of the debranching enzymes isoamylase and pullulanase, and the chain-shortening β-amylase, each with well-defined specificities, are used to elucidate structural features of amylose, amylopectin, and the intermediate fraction. The products of these treatments are then identified by chromatography (Fig. 6; Table III). Hizukuri (1986) observed that size-exclusion chromatography of the products of isoamylase action on amylopectin had a polymodal distribution (Fig. 6a); there are essentially five peaks (A, B1, B2, B3, and B4) with chain lengths as indicated. The model proposed by Hizukuri (Fig. 5b) takes into account this information, as the polymodal distribution in the chromatogram supports his idea of a cluster structure: 80 to 90% of the chains (A + B1) span a single cluster, about 10% (B2) would span (and connect) two clusters, 1 to 3% would span three clusters, and only 0.1 to 0.6% would connect four or more clusters. High-performance anion chromatograhy (HPAC) is another methodology that has proven to be a useful and sensitive tool for studying the structure of the linear chains released by debranching amylopectin and related carbohydrates (Fig. 6b).

The adjacent branch structures in amylopectin would form double helices that are organized in a crystalline structure (see preceding), provided that the various chains are of suitable length.

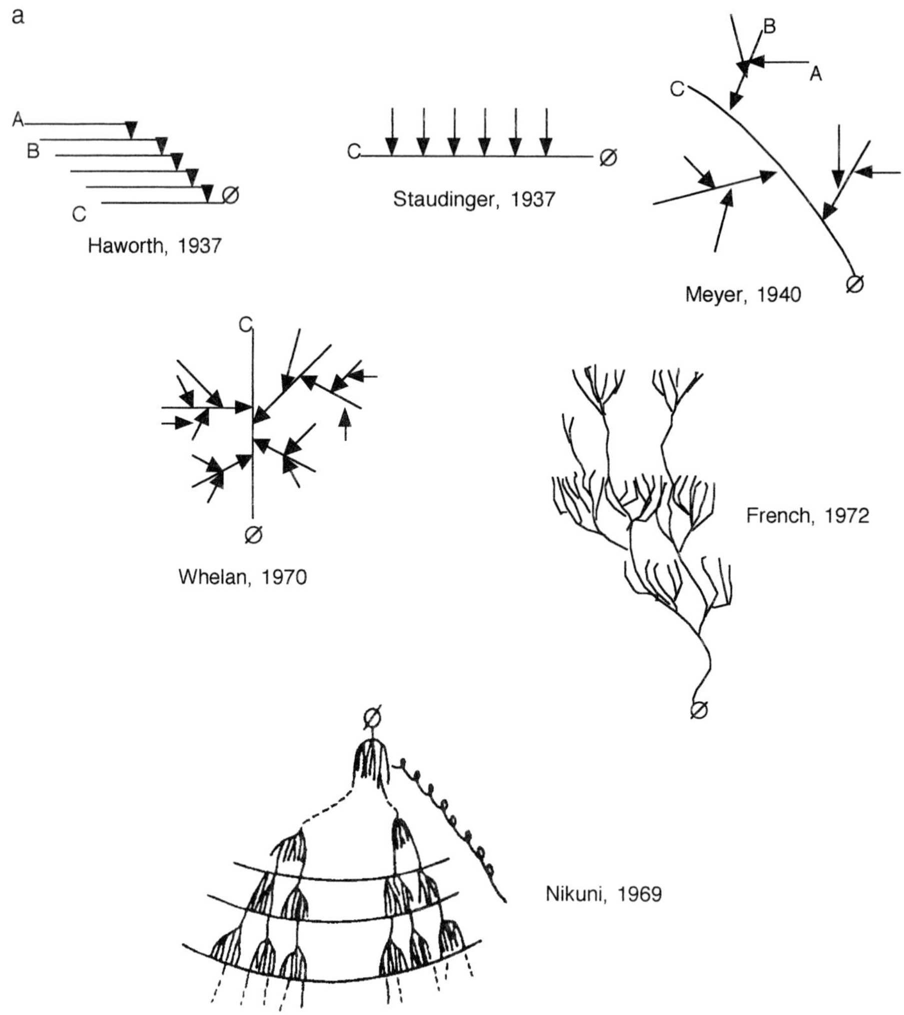

FIG. 5. (a) Historical evolution of the models for the structure of amylopectin as proposed by several workers; what varies in each model is the arrangement of the linear α-(1,4)-glucan chains and how they are joined by α-(1,6)-glycosidic linkages (arrowheads). (b) The model of Hizukuri (1986) showing A-, B_1-, B_2-, and B_3-chains (the very long B_4-chains are not illustrated) is the one more broadly accepted. "A" indicates A-chains whereas "B1", "B2", and "B3" are the B-chains; the C-chain has the only reducing end group, Ø, in the polysaccharide. The B3-chains are longer than the B2-chains, which are longer than the B1-chains. The B2-, B3-, and B4-chains extend into 2, 3, and 4 cluster regions, respectively. The average chain lengths are 19 for B1, 41 for B2, 69 for B3, and 104 for B4. The shortest chain length is for the A-chains, which have no branch points.

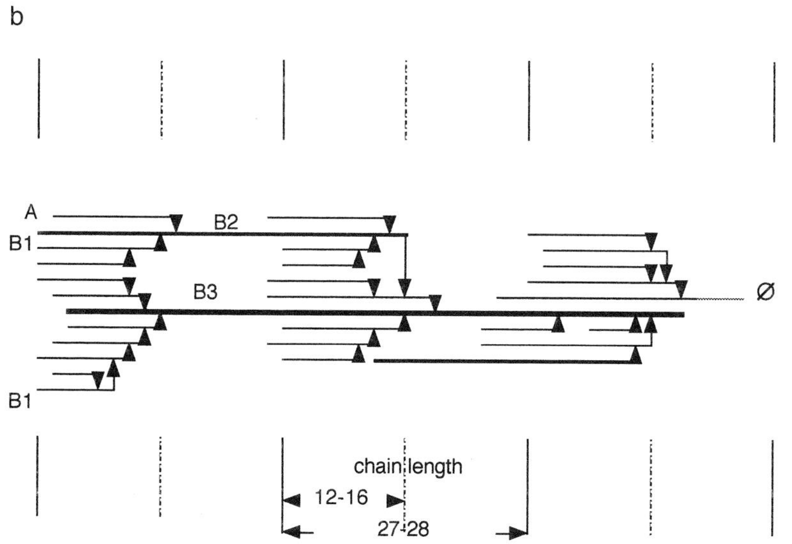

FIG. 5. (*Continued*).

The linear chains in the amylopectin form red to purple polyiodide complexes (Krisman, 1972) with a λ_{max} of between 530 and 585 nm. Altogether, the iodine binding capacity of amylopectin is very low, varying between 0 and 2.5 g/g depending on the botanical source of the amylopectin (Table IV). There are different kinds of atypical (anomalous) amylopectins (Baba *et al.,* 1987; Hizukuri, 1986; Takeda and Hizukuri, 1987), but they all bind more iodine and give a higher λ_{max} with I_2/KI solutions, leading to errors in determining the amylose content in starch when using the blue value (BV) or iodine affinity (IA) in the calculations. The IA is measured by amperometric titration; as iodine is added, the electric current does not increase until all the amylose molecules are saturated with iodine. Conversely, amylopectin cannot easily form the helical complex because the short chains and many branch linkages interfere with its formation. The BV is the absorbance at 680 nm of the iodine–glucan complex, under defined conditions, and can also be used to calculate the approximate proportion of amylose and amylopectin. One of the factors that affects the reliability of the IA and the BV as indicators of the proportion of amylose in the starch is the presence of lipids (relatively high in cereals), which also bind iodine.

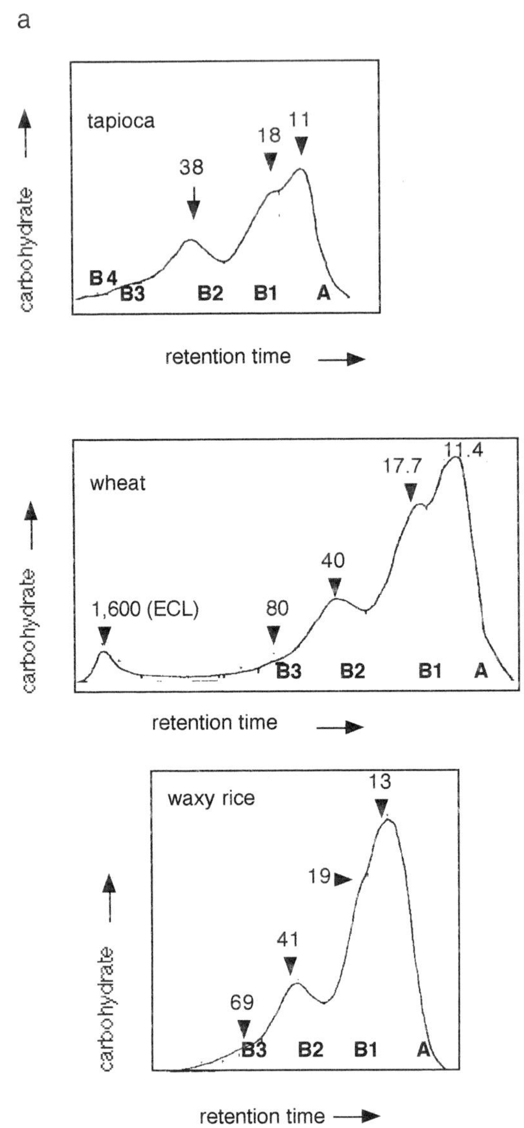

FIG. 6. (a) Size-exclusion high-performance liquid chromatography of amylopectins after debranching by isoamylase, showing the different chain length distributions for amylopectin from different species. The lower the retention time, the longer the debranched side chain. After Hizukuri (1995). (b) High-performance ion-exchange chromatography (using pulsed amperometric detection) of the linear chains obtained by debranching of amylopectin using isoamylase. The numbers indicate the degree of polymerization of the linear chains, and the height of the peak the relative amount of each chain length within the amylopectin (i.e., chain length distribution). The lower the retention time, the shorter the side chain. After Koizumi *et al.* (1991).

b

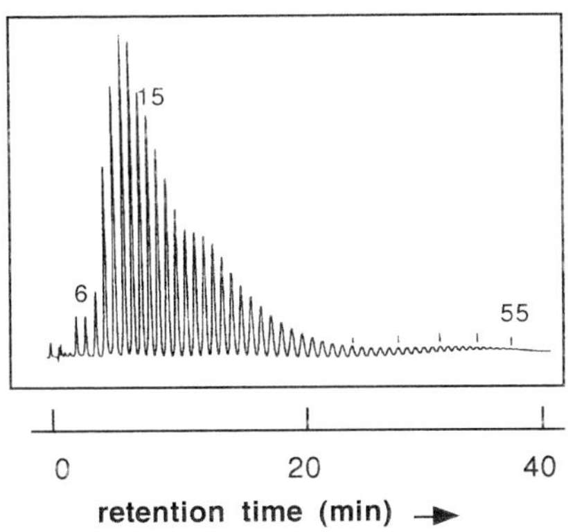

FIG. 6. (*Continued*).

III. MOLECULAR ORIENTATION IN THE GRANULE

Several levels of structural organization exist within the starch granule, as shown by the use of different methodologies. For example, starch granules show birefringence patterns in plane-polarized light that resemble maltese crosses (Fig. 1). Birefringence indicates a great degree of order in the molecular orientation, a characteristic that is independent of crystallinity; that is, noncrystalline polymers can show birefringence if their long axes are oriented by applied stress. The analysis of starch birefringence indicates that the chain axis of the polysaccharide is radially arranged. The

TABLE III

GENERAL PROPERTIES OF AMYLOPECTINS FROM DIFFERENT SOURCES[a]

Botanical source	B value	Iodine affinity	λ_{max}	β-Amylolysis limit (%)	Chain length	$[\eta]$	P0 (ppm)	P6 (ppm)
Wheat	0.098	0.89	552	57	20	145	9	<0.1
Maize (wild type)	0.11	1.10	554	59	22	—	14	4
Amylomaize	0.427	3.60	573	61	30	—	110	54
Rice	0.049	0.39	535	59	20	137	11	11
Barley	0.090	0.73	540	60	20	147	—	—
Sweet potato	0.166	0.44	—	56	22	180	135	101
Tapioca	0.104	—	—	57	21	—	—	—
Potato	0.245	0.06	—	56	23	—	900	840

[a] Data from Hizukuri (1995).

TABLE IV

DISTRIBUTION OF CHAIN LENGTHS OF AMYLOPECTINS FROM DIFFERENT SOURCES[a]

Fraction	Whole	A	B1	B2	B3	B4	Long chain	A/B
Waxy rice								
Chain length (max)		13	19	41	69			
Weight (%)	100	50	26.2	18.9	4.1	0.8		
Mole (%)	100	69.2	21.7	8.0	1.0	0.1		2.2
Wheat								
Chain length (max)		11	18	40	80			
Weight (%)	100	42	32.7	16.7	3.2	0.9	4.5	
Mole (%)	100	63.2	28.4	7.5	0.8	0.1	0	1.7
Tapioca								
		11	18	38	62			
	100	38.5	32.5	23.0	5.1	0.9		
	100	59.6	28.7	10.2	1.4	0.1		1.5
Potato								
		16	19	45	74			
	100	27.8	34.9	26.0	9.1	2.3		
	100	44.2	38.1	14.0	3.1	0.6		0.79

[a] Data from Hizukuri (1995).

constraints imposed on the growth of the starch granule by the shape of the plastid in which it is formed determine its shape, but do not seem to affect the internal arrangement of the molecules, which is analogous to that shown by semicrystalline polymers growing in solution.

Optical microscopy, scanning electron microscopy, or transmission electron micrographs of etched and stained thin sections of starch granules show a layered "growth ring" structure with ring spacings on the order of 1 μm. These rings are particularly visible after chemical treatment, and they represent shells of higher and lower starch content produced by the rate or mode of starch deposition. According to French (1984), growth rings would represent periodic growth and, with the cereal starches, daily fluctuations in the amount of carbohydrate available for starch deposition. The arms of the polarization cross are always perpendicular to the growth rings, indicating that the optic axes of the starch crystallites are aligned perpendicularly to the growth rings and the granule surface.

It seems that the molecules of amylose and amylopectin are arranged radially within the granule, at a right angle with the surface, and with their hypothetical single reducing end group toward the hilum or center of the granule (see Chapter 7, Initiation of Starch Synthesis). Growth of the granule is by aposition at the outer nonreducing end.

Scattering and diffraction measurements coupled with electron microscopy have shown layering within the granule, with unit blocks spaced at regular intervals of about 10 nm. These would be the result of regular spacing between clusters of partially crystalline amylopectin branches (Manners, 1989; Oostergetel and van Bruggen, 1989).

The model proposed by Kainuma and French (1972) attempts to explain the structure of the granule in relation to the polysaccharides that constitute it. The model shows the direction of growth of the amylopectin molecule; the crystal would grow at a right angle to the length of the molecule. Striations in the granule are caused by alternations of crystalline and amorphous zones, and these striations can be better observed with scanning electron microscopy of sections of granules partially digested with amylases or by acid treatment, which preferentially attacks the amorphous regions. At even higher magnification, and using transmission electron microscopy, it is possible to distinguish heterogeneities that are of the same magnitude as the A- and B-chain segments of amylopectin that are capable of forming crystalline parallel-stranded double helices. Order of the crystallites is probably responsible for the birefringence mentioned previously.

IV. METHODOLOGY AND NOMENCLATURE USED IN STARCH ANALYSIS

Much of what we know about starch structure is the result of the painstaking research of the pioneers of polysaccharide chemistry in the 1930s and 1940s: W. N. Haworth, E. L. Hirst, D. J. Bell, E. G. V. Percival, and others, who by using periodate oxidation, methylation, and paper chromatography established the basis for others, such as D. French, W. J. Whelan, and D. J. Manners to build on in the 1950s and 1960s. The purification and characterization of amylolytic enzymes from several sources have made possible the use of enzymatic, rather than chemical, identification of the starch components. Why is a detailed identification of the starch components so important? When a mutation affecting a particular enzyme results in changes in the seed appearance, the resulting changes in starch structure may be subtle: for example, a slight decrease in the average chain length of amylopectin or a small increase in the proportion of amylose to amylopectin. The resulting changes in the pasting quality of the starch, and/or its temperature of gelatinization, may be important to the industrial users. Identification of the enzymatic changes and of the consequent modification of the starch formed will help to illuminate the process of starch biosynthesis and to facilitate the task of "designing" a raw material that matches the needs of the industrial user.

V. OTHER CONSTITUENTS OF THE STARCH GRANULE

Although the proportion of amylose and amylopectin and their properties are paramount in determining the characteristics of the starch, minor constituents of the starch granule seem to affect the properties relevant to its use as food and in industrial applications. These minor constituents are not just contaminants (e.g., particles of bran that remained after scraping the wheat grain from the outer bran layer), but materials that are associated with the surface of the grain or are true internal components. The surface components may be remainders of the amyloplast in which the starch grain was formed (during the grain maturation, the amyloplast envelopes are disrupted and may remain on the surface of the granule), or the components may be endosperm proteins that became strongly attached to the granule during the maturation and drying of the grain. The surface components can be washed with water or salt solutions. Conversely, the internal components are part of the granule, and to extract them the starch granule must be disrupted.

VI. LIPIDS

Cereal starches contain low levels of lipids (0.5–1%), which are generally polar lipids requiring polar solvents such as methanol–water for extraction. Lipid content increases with amylose content, and unless the granule integrity is disrupted, the lipids remain inaccessible to normal fat solvents, suggesting that they are present as an amylose inclusion complex. Noncereal starches contain essentially no lipids.

Starches contain phosphorus, nitrogen, and very low amounts of other minerals. In the cereals, most of the phosphorus is in the form of phospholipids, whereas in potato starch the phosphorus is esterified to certain glucose residues in the polysaccharide.

VII. PHOSPHORUS

Generally, the phosphorus content in starches is associated with different pasting properties, and it confers a larger ion binding capacity. In wheat and corn starch, phosphorus is present largely or wholly as adsorbed phosphatides (extractable with boiling 85% methanol) associated preferentially with the amylose fraction.

Many amylopectins, but not amyloses, contain small amounts of esterified phosphate groups, present as residues of glucose 6-phosphate. Adsorbed phosphatides can be removed with suitable solvents, but esterified phosphate, such as that present in potato amylopectin, remains. The content of esterified phosphate varies between 200 and 1000 ppm in potato amylopectin and 40 and 150 ppm in starch from other tubers and roots but is very low in cereal starch (less than 20 ppm), with the exception of amylomaize, which contains 110–260 ppm (Takeda *et al.*, 1993).

Esterified phosphates have a marked effect on the physical properties of amylopectin and the extent of degradation by α- and β-amylase. After amylolysis, phosphorus is concentrated into the limit dextrins. The origin of the esterified phosphate, which would be close to the branching points, is not known. In general, glucose 6-phosphate and other glycolytic intermediates are not substrates for starch-synthesizing enzymes, and it still remains to be determined whether the glucose 6-phosphate residues are incorporated into a growing $(1\rightarrow4)$—α-D-glucan chain or arise from enzymatic phosphorylation of certain residues in amylopectin. Whatever the mechanism, it is not random, since one-third of the phosphate groups are present in the inner regions of B-chains and two-thirds are present in the outer parts of the B-chains and in the A-chains.

The position of the esterified phosphates at C-6 can be determined using acid hydrolysis followed by enzymatic methods, but ^{31}P NMR has been used to determine the position of ester linkages at other positions, such as C-3 and C-2 (Lim and Seib, 1993; Kasemsuwan and Jane, 1994).

Phosphorus content of potato starch (Geddes *et al.*, 1965) varies during development of the tubers; it increases from being undetectable in the amylose of tubers of 1-cm diameter, to 0.005% in tubers 8 to 9 cm in diameter, and, in amylopectin, from 0.029% in the 1-cm diameter to 0.049% in the largest (8- to 9-cm tubers).

Total phosphorus is determined as inorganic phosphate after treatment with hot perchloric acid. Phosphorus in D-glucose-6-phosphate residues is assayed using D-glucose-6-phosphate dehydrogenase.

VIII. PROTEINS

It has been known for many years that even after exhaustive washes, starch still contains small amounts of noncarbohydrate elements. Lipids account for most of the phosphorus and about one-third of the nitrogen present in wheat starch (Table I). However, amino acids have been recovered from hydrolyzed starch, indicating that the balance of nitrogen is present as proteins. In the case of well-isolated, nondamaged wheat starch, proteins represent 0.15 to 0.2% of its weight. This is a very small proportion of the flour protein, which comprises all the proteins contained in the endosperm, including storage proteins. However, small, granule-bound proteins have attracted attention for several reasons. Some, if not all, of the starch proteins are likely to be implicated in the formation of the granule. The drastic methodology required to extract these proteins from the granule makes their purification and characterization a difficult task. Some information is available on the waxy protein (see the chapter, "Starch Synthases") and on the pollen starch protein implicated in human allergy.

Proteins affect the milling and baking properties of the starch (Gough *et al.,* 1985; Greenwell *et al.,* 1985); the presence of one polypeptide in particular, of Molecular Weight 15,000, seems to determine the degree of "hardness" of wheat endosperm.

Inhalation of pollen present in the air provokes IgE-mediated responses of hay fever and allergic asthma in about 20% of humans. The allergens present in the pollen of rye grass (*Lolium perenne,* one of the grasses implicated in this response and the one that produces the greatest amount of pollen) are a group of low-molecular-weight proteins (Singh *et al.,* 1990). Electron microscopy shows that one of them, *Lolp*Ib, of Molecular Weight 31,000 and isoelectric point 9.0, is associated with the starch granules. *Lolp*Ib accumulates as the pollen grain matures, but its physiologic role is unknown. On contact with water the pollen grains burst, releasing the starch granules (about 1000 per grain), which are small enough to pass the barriers present in the mucose membranes, and amplifying the allergic response (Singh *et al.,* 1991).

FURTHER READINGS

These sources provide additional in-depth coverage of this topic. For complete reference, please see the Reference section at the end of the book.

Banks, W., and Greenwood, C. T. (1975)
Banks, W., and Muir, D. D. (1980)
Fitt, L. E., and Snyder, E. M. (1984)
French, D. (1984)
Hizukuri, S. (1995)
Imberty, A., Buléon, A., Tran, V., and Perez, S. (1991)
Kainuma, K. (1988)
Manners, D. J. (1985)
Morrison, W. R., and Karkalas, J. (1990)
Whistler, R. L. (1964)

BIOSYNTHETIC REACTIONS OF STARCH SYNTHESIS

I. INTRODUCTION

The metabolic routes leading to polyglucan synthesis were elucidated after the discovery of nucleoside-diphosphate sugars by L. F. Leloir and co-workers in 1955. This finding led to the conclusion that biosynthesis and degradation of glycogen and starch occur by different pathways.

In mammalian cells, glycogen synthesis is relatively well understood; glycogen synthase is specific for UDPglucose and is regulated through hormonally induced posttranslational protein modification. Textbooks of biochemistry usually describe these metabolite schemes in detail. The biosynthesis of polysaccharides in bacteria and plants is, in contrast, usually described in less detail. These organisms accumulate glycogen (bacteria) or starch (plants) by metabolic pathways that are different than those occurring in animals. Despite the difference in the final product (glycogen or starch), in bacteria and in plants ADPglucose is the glucosyl donor for the elongation of the α-1,4-glucosidic chain. Moreover, in both plants and bacteria, the main regulatory step of the metabolism takes place at the level of ADPglucose synthesis.

II. PIONEERING STUDIES

The formation of α-1,4-glucosidic linkages in vitro by plant enzymes was first demonstrated in 1940, when Hanes showed that potato tuber extracts formed an amylose-like product with glucose-1-P as a glucosyl donor (reaction 1).

$$\text{glucose-1-P} + \alpha\text{-glucan primer} \Leftrightarrow \text{Pi} + (1\rightarrow 4)\text{-}\alpha\text{-glucosyl-glucan} \quad (1)$$

Since then, phosphorylase activity has been found to be ubiquitous in plant extracts, but its role in vivo is now believed to be starch degradation rather than synthesis.

In the 1960s, L. F. Leloir, C. E. Cardini, and their collaborators in Buenos Aires, Argentina, demonstrated the synthesis of α-1,4-glucosidic linkages by a plant extract using either UDPglucose (Leloir *et al.,* 1961) or ADPglucose (Recondo and Leloir, 1961; Frydman and Cardini, 1967) as glucosyl donors (reaction 2), which is catalyzed by the starch synthase (EC 2.4.1.21).

$$\text{ADP(UDP)glucose} + \alpha\text{-glucan} \rightarrow \text{ADP(UDP)} + (1{\rightarrow}4)\text{-}\alpha\text{-glucosyl-glucan} \quad (2)$$

Since then, starch synthase activity has been reported to be present in many plant extracts (for reviews, see Preiss and Levi, 1980; Preiss and Sivak, 1996).

The sugar nucleotides UDPglucose and ADPglucose can be synthesized in plants either by a pyrophosphorylase-type reaction (reactions 3a and 3b; Espada, 1962) or via a reversal of the sucrose synthase reaction (reaction 4; Cardini *et al.,* 1955; de Fekete and Cardini, 1964)

$$\alpha\text{-glucose-1-P} + \text{ATP} \Longleftrightarrow \text{ADPGlc} + \text{PP}_i \quad (3a)$$

$$\alpha\text{-glucose-1-P} + \text{UTP} \Longleftrightarrow \text{UDPGlc} + \text{PP}_i \quad (3b)$$

$$\text{sucrose} + \text{ADP(UDP)} \Longleftrightarrow \text{fructose} + \text{ADP(UDP)glucose} \quad (4)$$

Formation of the α-(1→6) linkage branch points present in amylopectin and phytoglycogen is catalyzed by the branching enzyme (EC 2.4.1.18; Bourne and Peat, 1945; Hobson *et al.,* 1950), also called the Q enzyme.

$$\text{linear glucosyl chain of } \alpha\text{-glucan} \rightarrow \text{branched chain of } \alpha\text{-glucan with } \alpha\text{-1}{\rightarrow}\text{6-linkage branch points} \quad (5)$$

III. THE ADPglucose PATHWAY IS THE MAJOR PATHWAY OF STARCH SYNTHESIS *in Vivo*

Which of the enzymatic activities mentioned previously are involved in starch synthesis *in vivo?* To accept that an enzyme is a likely component of the pathway in the plant itself, it must fulfill the following criteria:

1. On careful extraction (i.e., avoiding proteolysis, inactivation by phenolics), from the plant tissue, the maximal activities measured *in vitro* (i.e., in the presence of activators at optimum pH) should at least equal the rates of starch synthesis measured *in vivo.*
2. The enzyme should be in the right compartment within the cell.
3. The kinetic characteristics (i.e., affinity for the substrate, effect of activators and inhibitors, pH optimum of the enzyme should be compared

with the concentrations of these substrates and modifiers in the site of starch synthesis (i.e., the chloroplast or amyloplast). This comparison will help in determining whether the activity in situ is likely to be sufficient to support the actual rate of synthesis measured in vivo. Calculation of the in vivo concentration of a particular metabolite is not an easy task; it involves the isolation of the organelle in question with minimum disruption and the avoidance of postisolation changes. With the nonaqueous techniques, the tissue is quickly frozen and the composition of the different components is assumed to be unchanged throughout the nonaqueous fractionation. The aqueous methods rely on fast separation of the different compartments with minimal cross-contamination.

4. Mutations resulting in the loss of a relevant enzyme should result in a commensurate decrease in starch content or a significant change in starch structure.

5. There should be a correlation between increases in the relevant enzymatic activities and the accumulation of starch during the development of the tissue (e.g., the potato tuber or the maize seed).

With these statements in mind, it is easier to address the reports proposing that UDPglucose-specific starch synthases and starch phosphorylases may be involved in starch synthesis. Their high K_m values for their substrates (UDPglucose and glucose-1-P, respectively), as compared to concentration in the relevant cellular compartments, argues against a significant role in starch biosynthesis. In addition, the synthesis of UDPglucose, at least in the starch-synthesizing plant tissues studied so far, occurs in the cytosol and not in the amyloplast, and no significant transport of UDPglucose into the plastid has been reported. Phosphorylase catalyzes an equilibrium reaction in cells that have P_i concentrations in excess of glucose-1-P, indicating that it plays a role in starch degradation rather than in synthesis.

Data from a number of genetic and biochemical studies indicate that the ADPglucose pathway, involving the reactions described in the preceding text is very important for starch synthesis. Mutants of maize endosperm *shrunken* 2 and *brittle* 2 (Tsai and Nelson, 1966; Dickinson and Preiss, 1969b), which are deficient in ADPglucose pyrophosphorylase (ADPGlc PPase) activity, are also deficient in starch. Smith *et al.* (1989) have shown that a pea line having recessive *rb* genes (genes controlling the level of ADPGlc PPase activity in developing pea embryos), containing 3–5% of the ADPGlc PPase activity, had only 38 to 72% of the starch found in the normal pea line. In *Arabidopsis thaliana,* Lin *et al.* (1988b) isolated a mutant containing less than 2% of the starch seen in the normal strain and less than 2% of the ADPGlc PPase activity. Immunoblots indicated that the enzyme was absent from the *Arabidopsis* extracts. In the potato tuber,

Müller-Röber and colleagues (1992) expressed a chimeric gene encoding antisense RNA for the ADPGlc PPase small subunit, which caused a reduction in enzymatic activity of 2 to 5% of the normal levels, which led to a reduction in starch content.

Thus, in four different plant systems, a reduction of ADPGlc PPase activity led to a reduction in starch accumulation. Alternatively, an increase in ADPGlc PPase activity was achieved by transformation of the potato tuber with a mutant *E. coli* ADPGlc PPase gene that was insensitive to the regulatory effectors of the plant enzyme (Stark *et al.,* 1992). This increased the potato tuber starch content by 30 to 60%, suggesting not only that the role of the ADPGlc PPase in starch synthesis is important, but also that the enzyme activity is normally rate limiting. The introduction of the bacterial gene into tomato fruit (Stark *et al.,* 1992) and into safflower seed (G. Kishore, personal communication, 1997) also increased their starch content dramatically. Other data showing a relationship between activity of the ADPGlc PPase and starch accumulation in other plant species have been previously reviewed (Preiss and Levi, 1980; Preiss, 1988, 1991; Okita, 1992; Sivak and Preiss, 1995; Preiss and Sivak, 1996). Thus, the ADPGlc PPase and the subsequent reactions utilizing ADPglucose are the dominant routes for starch synthesis in plants, and ADPglucose synthesis is perhaps rate limiting. In the case of starch phosphorylase, the first criterion is fulfilled, but the concentrations of P_i and glucose-1-P in the amyloplast and chloroplast are considered to be more compatible with a role of the enzyme in degradation rather than in synthesis. No correlation between plastid phosphorylase activity and starch accumulation has been found. No mutants deficient in starch synthesis have been found that are deficient in phosphorylase. Some reports suggest that phosphorylase may play some role in starch synthesis (Obata-Sasamoto and Suzuki, 1979; Mengel and Judel, 1981), a conclusion based on the fact that phosphorylase levels were higher than starch synthase and/or ADPGlc PPase. It should be noted, however, that insensitive assays for the ADPglucose enzymes were frequently used, and that although phosphorylase was found to be higher in activity, the physiologic concentrations of P_i and glucose-1-P make it unlikely to ever function in a synthetic pathway. The equilibrium constant for phosphorylase is 2.4 at pH 7.3 (Cohn, 1961). The ratio of P_i to glucose-1-P has been estimated at about 3 : 300 (Heber and Santarius, 1965; Bassham and Krause, 1969), and subsequent studies agree with this ratio. The K_m values measured for glucose-1-P are one to two orders of magnitude higher than the glucose-1-P concentration calculated for the whole cell. Thus, although the phosphorylase activity, when tested at saturating concentrations of substrate, appears to be higher than starch synthase, in physiologic conditions this may not be the case. Still, it is possible that in conditions favoring starch synthesis, and in the site of starch synthesis (the chloroplast or amyloplast), concentrations of P_i may be lower and concentrations of glucose-1-P may be higher than the whole-cell concentrations averaged over time.

One approach that would be useful in finding the physiologic role(s) of the starch phosphorylase would be the expression of antisense RNA in an organ such as the potato tuber [as Müller-Röber and colleagues (1992) have done for the AdPGlc PPase; see the following], followed by measurement of the corresponding enzyme (plastidial or cytosolic), and a thorough study of the effects (if any) of the consequence deficiency of the amyloplast phosphorylase on the amount and structure of the starch formed.

IV. ALTERNATIVE PATHWAYS

In studies reviewed previously (Preiss and Levi, 1980; Preiss, 1988), no relationship was noted between starch synthesis and UDPglucose pyrophosphorylase (UDPGlc PPase) (reaction 3b) activity. The high K_m values of starch synthase for UDPglucose, as compared to the measured cellular levels, strongly argue against a significant role for UDPGlc PPase in starch synthesis. It would seem that the major part, if not all, of UDPGlc PPase activity is localized in the cytosol and not in the organelle involved in starch synthesis—the amyloplast (Bird *et al.,* 1974; Robinson and Walker, 1979; Macdonald and ap Rees, 1983). Thus, this high activity does not appear to be localized where active starch synthesis occurs. In contrast, the maximum activities of ADPGlc PPase and ADPglucose-starch synthase, which are localized in the amyloplasts, are at least three times greater than the rate of starch accumulation in soybean cultures and 1.3 to 2.7 times greater than the rate of starch accumulation in the developing club of the spadix of *Arum* maculatum (ap Rees *et al.,* 1984).

An alternate pathway for starch synthesis has been proposed, which is based on the finding of a putative ADPGlc translocator in the envelope of both amyloplasts and chloroplasts. Akazawa *et al.* (1991) proposed that ADPglucose is synthesized in the cytosol by the sucrose synthase (rather than in the plastid by the action of the ADPGlc PPase, as is widely accepted), and is then transported into the plastid where it is converted into starch by the starch synthase. A critique of this hypothesis is presented in the chapter, "The Site of Starch Synthesis in Nonphotosynthetic Plant Tissues: The Amyloplast," where metabolite transport into the plastids is discussed, but it is worth mentioning here that this pathway does not fulfill the criteria mentioned in the preceding—that is, the experimental evidence does not support this alternative pathway.

Conversely, a large body of evidence strongly indicates that the main, if not the only, pathway of starch synthesis consists of the enzymatic reactions catalyzed respectively by ADPGlc PPase (reaction 3a), the starch synthase reaction (reaction 2), and the branching enzyme (reaction 5). The data supporting this view are from a number of biochemical and genetic studies.

V. RATE OF STARCH SYNTHESIS VERSUS ACTIVITIES OF THE STARCH BIOSYNTHETIC ENZYMES

A direct relationship between the increase in the activities of starch synthase and ADPGlc PPase, and the rate of starch accumulation, has been reported for developing maize endosperm (Ozbun *et al.,* 1973), wheat grain (Moore and Turner, 1969; Turner, 1969), and potato tubers (Sowokinos, 1976) (for a review of these, see Preiss and Levi, 1980).

The man-made intergenic hybrid, triticale (X-*Triticosecale* Wittmarck), may produce in development either plump or shriveled seeds (Dedio *et al.,* 1975). The difference between the two seeds was originally postulated to be due to higher amylase content. However, studies in which the activities of the starch biosynthetic enzymes and amylase were measured during seed development showed that even though the shriveled seeds contained more amylase than the plump seeds, the shriveled appearance occurred earlier than the increase in amylase activity (Ching *et al.,* 1983). Starch synthase and ADPGlc PPase activities, extracted at different stages of the seed development, were in excess of the measured rates of starch accumulation, indicating that these enzymes could play an important role in starch synthesis.

Similar results were obtained with germinating seeds of *Ricinus communis* (Reibach and Benedict, 1982). Starch levels increased about two-fold in the imbibed seed in 5 days. The starch synthase and ADPGlc PPase activities increased to a maximum 4 to 5 days after germination and were high enough to account for the observed rates of starch synthesis. In this study, the ADPGlc PPase activity was 1000-fold higher than the previously reported UDPGlc PPase activity (Nishimura and Beevers, 1979).

In maize ears, the apical kernels develop and pollinate several days after basal kernels; the kernels that are formed earlier may have higher survival probability, longer growth duration, and higher growth rates. Ou-Lee and Setter (1985b) compared the activities of the starch biosynthetic enzymes in the apical and basal kernels during development of synchronously pollinated ears. During the period of maximal starch synthesis, the ADPGlc PPase and starch synthase activities could account for the observed starch accumulation rate in basal kernels, but were slightly less than adequate to account for starch synthesis in apical kernels. It should be pointed out, however, that it was later shown (Plaxton and Preiss, 1987) that during extraction, maize endosperm ADPGlc PPase is particularly sensitive to endogenous protease activity, which significantly changes the regulatory properties of the enzyme and decreases its stability. It is therefore possible, as Ou-Lee and Setter (1985b) suggested, that their assays may have underestimated the starch biosynthetic enzyme activities but, nevertheless, a rough correlation between maximal starch accumulation and the levels of the starch biosynthetic enzymes was noted. In an extension of this study, the effect of differential temperature increases on the growth rate and size of the apical kernels was examined (Ou-Lee and Setter, 1985a). The temperature was increased to 25°C at 7 days after pollination, as opposed to the lower temperatures normally experienced by the plant at nights and in cool weather. The tip-heated treatment slightly increased the size of the apical kernels at the expense of slightly decreasing the size of more numerous basal and middle position kernels, and some of the developmental events in the apical kernels were accelerated. The maximal levels of ADPGlc PPase and starch synthase activities occurred earlier and correlated well with the earlier rise of starch levels in the heated apical kernels, and the ADPGlc PPase activities were sufficient to account for the starch content measured. Starch synthase activities, measured as granule-bound enzymes, were insufficient. Soluble starch synthase activity was not measured.

Addition of adenine (0.1 m*M*) to cultures containing tobacco callus cells increased the starch content almost 4-fold in 3 days (Gamanetz and Gamburg, 1981). Addition of other purine or pyrimidine bases had no effect on the starch content. In the cells grown with adenine, there was a 100-fold increase in the ADPglucose content (there was no effect on the UDPglucose content) and a 2.5-fold increase in the specific activity of ADPGlc PPase. Adenine has been shown to increase the starch content of other plant cells in suspension cultures (i.e., soybean, potato, *Atriplex* sp., dewberry) (Gamanetz and Gamburg, 1981), and it was concluded that in the plant cells the adenine pool may be limiting for ADPglucose synthesis. Addition of adenine would stimulate ADPglucose synthesis and, therefore, starch synthesis.

In a series of experiments to determine which sugar nucleotide—ADPglucose or UDPglucose—plays the major role in starch synthesis in nonphotosynthetic plant cells, ap Rees *et al.* (1984) estimated the in vivo rates of starch synthesis in the developing club of the spadix of *Arum maculatum* and in suspension cultures of soybean. They compared these estimates with the maximum catalytic activities of four enzymes: ADPglucose- and UDPglucose–starch synthase, ADPglucose pyrophosphorylase, and UDPglucose pyrophosphorylase; the amounts of ADPglucose and UDPglucose in these cells were also determined. The conclusion was that in *Arum* clubs and soybean cultures, starch synthesis proceeds almost entirely via ADPglucose.

VI. A MISSING STEP?

As discussed previously, only three reactions—those catalyzed by ADP-Glc PPase, starch synthase, and branching enzyme—are needed to synthesize all the glucosidic linkages found in the starch granule. However, it is interesting to note that the *sugary* (*su*) 1 mutation in the maize endosperm does not affect the expression of the genes of any of the three activities, but still results in a significant reduction in starch granule formation. This decrease in starch accumulation is accompanied by an increase in the content and an increase in the content of a water-soluble α-1,4-glucan phytoglycogen in such a way that the total polysaccharide content approaches that of normal maize. Thus, another enzyme activity may be required to complete the formation of the starch granule. Pan and Nelson (1984) showed that maize endosperm displaying the *su* 1 mutation was defective in debranching enzyme activity. More recently, the *su* 1 was cloned (James *et al.,* 1995), and sequence analysis of its cDNA showed that it has a high degree of homology with a bacterial isoamylase (Yang *et al.,* 1996). Thus, Ball *et al.* (1996) proposed that the *su* 1 gene, believed to be the structural gene for isoamylase activity, is required for formation of the finished amylopectin product. In other words, a fourth enzyme would be needed to convert the product of the branching enzymes into amylopectin, which is able to crystallize, trapping the amorphous amylose to form the starch granule. This subject is discussed more extensively in the chapter, "Open Questions and Hypotheses in Starch."

VII. SUMMARY

The major route to starch biosynthesis involves three reactions. The first reaction, catalyzed by ADPGlc PPase (glucose-1-P adenylyltransferase; EC 2.7.7.27), results in the synthesis of the glucosyl donor ADPglucose. The second reaction, catalyzed by starch synthase (ADPglucose: 1,4-α-D-glucan 4-α-D-glucosyltransferase; EC 2.4.1.21), transfers the glucosyl group of ADPglucose to the nonreducing end of an α-1,4-glucan primer to form a new α-1,4-glucosidic bond. The synthesis of the α-1,6-branch linkages found in amylopectin is catalyzed by branching enzyme (1,4-α-D-glucan: 1-4-α-D-glucan 6-glycosyl-transferase; EC 2.4.1.18).

The kinetic properties of the enzymes in the ADPglucose pathway (K_m and V_{max} values), together with the concentrations of substrate and effector metabolites in plant cells, are consistent with a major role for the pathway in starch synthesis. Conversely, the properties of the UDPglucose-specific starch synthases and starch phosphorylases (i.e., the high K_m values for their substrates, UDPglucose and glucose-1-P, respectively), as compared to the concentration in the relevant cellular compartments, argue against a significant role or UDPglucose starch synthase and starch phosphorylase in starch biosynthesis. No relationship has been observed between starch synthesis and the activities of starch phosphorylase or UDPGlc PPase in the tissues studied. Moreover, in some starch-synthesizing plant tissues, the synthesis of UDPglucose only occurs in the cytosol and not in the amyloplast, where starch is made.

Analyses of the starch biosynthetic system in a number of plants and green algae indicate that an important site of regulation of starch synthesis is at the ADPGlc PPase and that 3PGA and P_i are important regulatory metabolites of that enzyme.

SYNTHESIS OF THE GLUCOSYL DONOR: ADPglucose PYROPHOSPHORYLASE

I. REGULATORY PROPERTIES

In the biosynthesis of starch and bacterial glycogen, the glucose donor, ADPglucose, is formed from ATP and glucose-1-P via a reaction catalyzed by ADPglucose pyrophosphorylase (ADPGlc PPase; glucose-1-P adenylyltransferase; E.C. 2.7.7.27). This reaction was first described by Espada (1962) in soybean and was subsequently found in many plant tissues and in bacterial extracts. ADPGlc PPases have been isolated from many plants and bacteria, and their regulatory properties have been studied. Although the major activators vary according to the source, they share a common characteristic: The activator specificity of the enzyme is determined by the major pathway of carbon assimilation in the organism. The relationship of activator specificity of the ADPGlc PPase of the various organisms with the pathways is summarized in Table I.

The reaction catalyzed by ADPGlc PPase is reversible, and it should be noted that regulatory properties can be different in the two directions. For example, pyrophosphorolysis is usually much less affected by allosteric activators than is the synthesis of the sugar nucleotide (e.g., see Ghosh and Preiss, 1966; Preiss *et al.,* 1967).

Enteric bacteria, such as *Escherichia coli,* assimilate glucose via glycolysis and regulation of the glycolytic pathway is at the site of fructose-1,6-bis-P synthesis (the phosphofructokinase step), and this is the major activator for the *E. coli* ADPGlc PPase (Preiss, 1984; Preiss and Romeo, 1989, 1994).

For organisms where the predominant pathway is the Entner–Doudoroff pathway, fructose-1,6-bis-P is not a major metabolite in glucose degradation (because glucose-6-P is converted first into 6-P-gluconate and then to 2-keto,3-deoxy,6-P-gluconate); the activators for their ADPGlc PPase are fructose-6-P and pyruvate (Preiss, 1969, 1984; Preiss and Romeo, 1989).

Rhodospirillum rubrum cannot metabolize glucose but grows anaerobically on pyruvate, lactate, or on CO_2. Pyruvate has been shown to be a product of CO_2 fixation, and it is also the sole activator of the *R. rubrum* ADPGlc PPase (Furlong and Preiss, 1969).

TABLE I

ACTIVATOR SPECIFICITIES OF ADPGLUCOSE PYROPHOSPHORYLASES (ADPGlc PPASE) FROM DIFFERENT ORGANISMS[a]

Organisms	Activator specificity	Assimilation pathway
Enterobacteria	Fructose-1,6-bis-P	Glycolysis
Agrobacterium tumefaciens	Fructose-6-P, pyruvate	Entner–Doudoroff pathway
Rhodopseudobacter spheroides	Fructose-1,6-bis-P, pyruvate, fructose-6-P	Glycolysis, Entner–Doudoroff, anaerobic photosynthesis
Rhodospirillum rubrum	Pyruvate	Anaerobic photosynthesis
Cyanobacteria, green algae, higher plants	3-P-Glycerate	Oxygenic photosynthesis

[a]Grouped according to carbon assimilation pathway.

Rhodobacter spheroides, a highly adaptable organism, can metabolize glucose by glycolysis or, under other physiologic conditions, by the Entner–Doudoroff pathway, and it can also assimilate CO_2 during anaerobic photosynthesis. It has an ADPGlc PPase that is effectively activated either by fructose-1,6-bis-P, fructose-6-P, or pyruvate (Greenberg *et al.,* 1983) (i.e., its adaptability in carbon assimilation is associated with an ADPGlc PPase with flexible activation specificity).

Cyanobacteria, green algae, and higher plants assimilate CO_2 during photosynthesis to form 3-P-glycerate (3PGA). By 1982, ADPGlc PPases from several plant species—13 from leaf and 9 from nonphotosynthetic tissues—had been shown to be activated by 3PGA (Preiss, 1982b), which in most cases increases the affinity for the substrates, ATP and glucose-1-P, and reverses the inhibition caused by P_i. Since 1982, ADPGlc PPases from other nonphotosynthetic tissues have been studied (e.g., maize endosperm, potato tuber, cassava root, rice endosperm), and these tissues were highly dependent on the presence of 3PGA and were inhibited by P_i. Some exceptions to this rule have been reported. In the ADPGlc PPases from pea embryos (Hylton and Smith, 1992), barley endosperm (Kleczkowski *et al.,* 1993), and bean cotyledon (Weber *et al.,* 1995), activation by 3PGA is not as high, ranging between 1.5- and 3-fold. However, ADPGlc PPases are usually much less affected by allosteric activators in the pyrophosphorolysis direction than in the synthesis direction (Ghosh and Preiss, 1966; Preiss *et al.,* 1967). Activation for the "anomalous" enzymes would likely be higher if assayed in the synthesis direction, which is, after all, the direction in which the glucose donor is formed.

In the first studies of maize endosperm ADPGlc PPase, it was thought that the enzyme was insensitive to 3PGA activation and P_i inhibition (Dickinson and Preiss, 1969a,b). It was found later, however, that if protease inhibitors were added to the maize endosperm extracts, activity was then very sensitive to activation by 3PGA and to inhibition by P_i (Plaxton and Preiss, 1987). It was also shown that if the activity of proteases was not prevented, the size of the 54-kDa subunit was reduced to 53 kDa, a small but reproducible change in size. Thus, partial proteolysis during enzyme isolation can strongly affect ADPGlc PPase regulatory properties, and proteolysis may be one reason behind the allosteric insensitivity found in the atypical ADPGlc PPases.

Figure 1 illustrates how relatively small changes in the 3PGA and P_i concentrations can greatly affect the rate of ADPglucose synthesis, particularly at low concentrations of 3PGA, where the activation is minimal, and in the presence of P_i. At 1.2 m*M* P_i and 0.2 m*M* 3PGA, ADPglucose synthesis is inhibited by more than 95%. However, if the P_i concentration decreases 33% to 0.8 m*M*, and the 3PGA concentration increases 50% to 0.3 m*M*, there is an 8.5-fold increase in the rate of ADPglucose synthesis. Conversely, at 0.4 m*M* 3PGA and 0.8 m*M* P_i, the rate of ADPglucose

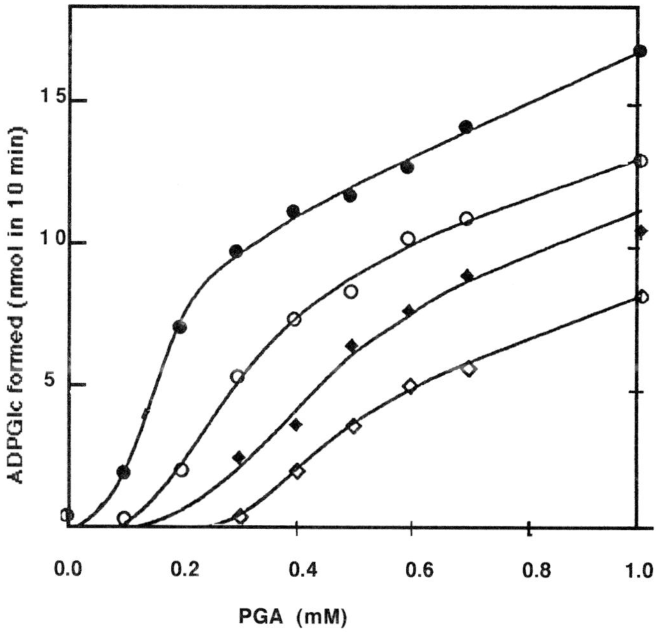

FIG. 1. Effects of P_i and 3PGA on rate of ADPGlucose synthesis catalyzed by potato tuber ADPGlc PPase. ●, 3PGA curve done in the presence of 0.4 m*M* P_i; ○, 0.8 m*M* P_i; □, 1.2 m*M* P_i; ■, 3PGA curve measured in the presence of 1.6 m*M* P_i.

synthesis is 7.5 nmol per 10 minutes. This is reduced to 2.2 nmol (70% decrease) if the 3PGA concentration decreases 50%, to 0.2 m*M*. If the P_i concentration increases to 1.2 m*M*, the synthetic rate is then reduced to 0.65 nmol, which is a reduction in ADPglucose synthesis of 91%. The reason that small changes in the effector concentrations produce such large effects in the synthetic rate is due to the sigmoidal nature of the curves particularly at the low concentrations of 3PGA.

II. PHYSIOLOGIC RELEVANCE OF THE ADPGlc PPase REGULATORY PROPERTIES

In vivo and in situ experiments strongly indicate that the activation by 3PGA and inhibition by P_i observed in vitro are also physiologically important. Many experiments have been cited in reviews (Preiss and Levi, 1980; Preiss, 1982a,b, 1988, 1991, 1996; Sivak and Preiss, 1995; Preiss and Sivak, 1996) showing a direct correlation between the concentration of 3PGA and starch accumulation, and an inverse one between P_i concentration and starch content. This is true for photosynthetic tissues, in which P_i and PGA concentrations within the chloroplast are good indicators of the energy and carbon status, and in this way the ADPGlc PPase provides a good regulatory mechanism for the flux of photosynthate into starch. It has been found that the regulatory properties of the enzyme of nonphotosynthetic tissue, such as potato tuber and maize endosperm, are such that the ADPGlc PPase is almost completely dependent on the presence of the activator, but in these tissues it is still uncertain how 3PGA and P_i can signal the availability of carbon and energy for starch synthesis, since transport of carbon in the amyloplast is via hexose-phosphates rather than by triose-phosphates as seen in chloroplasts (Keeling *et al.,* 1988; Heldt *et al.,* 1991; Hill and Smith, 1991; Viola *et al.,* 1991).

If this activation mechanism is indeed important physiologically, its failure should have important consequences in vivo. This has been confirmed by chemical mutagenesis in bacteria (Preiss, 1969, 1984, 1996), *Arabidopsis thaliana* (Lin *et al.,* 1988a,b), and in the green algae *Chlamydomonas reinhardtii* (Ball *et al.,* 1991). More recently, an allosterically altered ADPGlc PPase has been reported in maize endosperm (Giroux *et al.,* 1996). In the *Chlamydomonas* system, starch-deficient mutants have been isolated and characterized, and have been shown defective in the ADPGlc PPase, which could not be effectively activated by 3PGA. The maize endosperm ADPGlc PPase allosteric mutant is less sensitive to P_i inhibition than the normal enzyme and the mutant endosperm has 15% more dry weight than the normal endosperm (Giroux *et al.,* 1996). The *Chlamydomonas* starch-deficient and higher dry-weight maize endosperm mutants ADPGlc PPases strongly suggest that the in vitro regulatory effects observed with the photosynthetic and nonphotosynthetic plant ADPGlc PPases are highly functional in vivo, and that ADPGlc synthesis is rate limiting for starch synthesis.

III. SUBUNIT STRUCTURE

To study subunit structure, it is essential to determine the molecular mass of the holoenzyme by gel filtration and/or sucrose density gradient followed by determination of enzymatic activity. The size of the subunits can be determined by sodium dodecyl phosphate–polyacrylamide gel electrophoresis (SDS–PAGE). Put together, this information will show whether the enzyme is a monomer or a polymer and, if the latter, how many subunits make up the holoenzyme and whether there is only one kind of subunit or more than one kind.

Many bacterial ADPGlc PPases have been purified and in many their subunit structure has been determined. Invariably the native enzymes are tetrameric with only one kind of subunit, with a molecular mass ranging from 49,000 to 54,000, according to the species.

In contrast, the plant enzyme consists of two related but different subunits with masses in the 50,000 to 60,000 range. The "small" subunits have molecular masses of about 50,000 to 54,000, whereas the other, "large" subunits have molecular masses of 51,000 to 60,000. Although the difference in mass between the two subunits in one enzyme can be small, it is still convenient to designate them as small and large; they differ in many other characteristics, and this is discussed as follows.

The potato tuber, spinach leaf, and maize endosperm enzymes have small subunit masses of 50,000, 51,000, and 54,000, respectively, and large subunit masses of 51,000, 54,000, and 60,000, respectively. The small and large subunits have about 50 to 60% identity with each other and have about 30 to 40% identity with the procaryotic ADPGlc PPases.

An ADPGlc PPase that is well studied with respect to structural properties is the spinach leaf enzyme (Morell *et al.,* 1987, 1988; Ball and Preiss, 1994). This enzyme has a molecular mass of 206,000 and is composed of two different subunits, with molecular masses 51,000 and 54,000. These subunits, which can be separated by chromatography after denaturating the holoenzyme with urea, can be distinguished not only by their molecular masses but also with respect to amino acid composition, amino-terminal sequences, peptide patterns on high-performance liquid chromatography (HPLC) of their tryptic digests, and antigenic properties. The polyclonal antibody prepared against the 51-kDa subunit reacted very strongly, in immunoblots, with the 51,000 subunit, but weakly with the 54,000 subunit. Conversely, antibodies raised against the large subunit reacted only weakly with the small subunit and strongly with the large. Thus, on the basis of the protein chemistry and immunologic analyses, the two subunits are distinct and probably are the products of two genes.

Preiss *et al.* (1990) showed that the maize endosperm ADPGlc PPase, which has a molecular mass of 230,000, could react with the antibody prepared against the native spinach leaf enzyme in immunoblot experiments. In SDS gel electrophoresis of endosperm extracts or of the highly purified enzyme, two polypeptides of 55 and 60 kDa reacted with the antiserum raised against the spinach holoenzyme. The results were different when antibodies raised against the separate subunits (large or small) were used. The antibody prepared against the spinach leaf large subunit cross-reacted mainly with the endosperm large subunit and to a small extent with the 55-kDa subunit. The antibody against the spinach leaf small subunit antibody cross-reacted well with the endosperm 55-kDa subunit and weakly with the 60-kDa subunit. The maize endosperm starch-deficient mutants, *shrunken* 2 (*sh* 2) and *brittle* 2 (*bt* 2), were also studied. In immunoblotting experiments and while using antibodies against the native or subunit antibodies of the spinach leaf enzyme, the mutant *bt* 2 endosperm lacked the 55-kDa subunit and the mutant *sh* 2 endosperm lacks the 60-kDa subunit. These results indicate that the maize endosperm ADPGlc PPase is composed of two immunologically distinctive subunits, and that the *sh* 2 and *bt* 2 mutations cause reduction in ADPGlc PPase activity (and the consequent deficiency in starch content) through the lack of one of the subunits. Thus, the *sh* 2 gene would be the structural gene for the 60-kDa, large subunit, whereas the *bt* 2 gene would be the structural gene for the 55-kDa, small subunit.

An ADPGlc PPase cDNA clone, isolated from a maize endosperm library (Barton *et al.,* 1986), hybridized with the small subunit cDNA clone from rice (Anderson *et al.*, 1989). This maize ADPGlc PPase cDNA clone hybridizes to a transcript that is present in maize endosperm but absent in *bt* 2 endosperm. Thus, the *bt* 2 mutant appears to be the structural gene of the 55-kDa subunit of the ADPGlc PPase. These data also indicate that the nonphotosynthetic tissue ADPGlc PPase is also composed of two subunits and, on the basis of immunoreactivity, there is homology between the large and small subunits in the leaf enzyme with the subunits of a reserve tissue enzyme, respectively.

The potato tuber ADPGlc PPase has been highly purified and, by two-dimensional polyacrylamide gel electrophoresis, two polypeptides could be distinguished by their slight differences in molecular mass, 50,000 and 51,000, and in net charge (Okita *et al.,* 1990). The tuber small subunit is reactive with the antibody prepared against the spinach leaf small subunit. The antiserum prepared against the spinch leaf large subunit, however, does not react with either potato tuber enzyme subunit. The potato tuber enzyme is composed of two distinct subunits and is not a homomer as initially thought (Sowokinos and Preiss, 1982).

The ADPGlc PPase of *A. thaliana* is composed of two subunits, with molecular masses of 51,000 and 54,000. One *A. thaliana* mutant, TL25, lacks both subunits of the ADPGlc PPase (it is thought that the mutation affects a regulatory locus), whereas another mutant, TL46, lacks the large, 54-kDa subunit only. The TL46 mutation provides further evidence that the larger subunit is a necessary component of the native ADPGlc PPase for optimal activity since the mutant has only 7% of the wild-type activity. The mutant synthesizes starch at 9% of the rate displayed by the wild type in high light, and at 26% of the wild-type rate measured at low light (Neuhaus and Stitt, 1990).

IV. STRUCTURE–FUNCTION RELATIONSHIPS

The researcher who wants to elucidate the mechanism of action and the regulation of an enzyme has many methodological tools at his disposal, and more become available every year.

Chemical modification can supply information on the amino acids involved in the active and regulatory sites. The amino acid sequences obtained by Edman degradation of the proteins purified from different tissues and species, and/or by cloning followed by deduction of amino acid sequences, can be compared. This exercise will point out the amino acid sequences well conserved in enzymes from different sources, which are likely to be essential for enzyme function. Using site-directed mutagenesis, the amino acids deemed to be crucial are replaced by others, and the effect of these changes on the properties of the enzyme are studied. To achieve this objective, *E. coli* is transformed with the mutated gene in a suitable vector, the overexpressed enzyme is purified, and its properties are compared with those of the enzyme obtained from bacteria transformed with the nonmutated gene.

Chemical mutagenesis, followed by screening for starch with iodine reagent, can help identify amino acids crucial for binding or catalysis in an approach similar to that used for the ADPGlc PPase of *E. coli.* In plants, chemical mutagenesis has been used with *A. thaliana,* (Lin *et al.,* 1988a,b), with *C. reinhardtii* (Ball *et al.,* 1991), and with the potato enzyme expressed in *E. coli* (Greene *et al.,* 1996). As for any methodology intending to identify a crucial amino acid, the effect of the mutation in a single amino acid must be specific for a particular substrate or modulator. A generalized effect indicates that the amino acid in question affects the general conformation of the enzyme.

V. FUNCTION OF THE HIGHER PLANT ADPGlc PPase SUBUNITS

After discovering that the plant native ADPGlc PPases were tetrameric and composed of two different subunits, the next step was to determine why the two subunits were required for optimal catalytic activity. Since the enzyme must contain ligand binding sites for the activator (3PGA), inhibitor (P_i), sites for the two substrates (ATP and glucose-1-P), as well as a catalytic site, it is possible that these sites could be located on different subunits.

Two cDNAs encoding the mature large subunit and small subunits of the potato tuber (*Solanum tuberosum* L.) ADPGlc PPase have been expressed in *E. coli* (Iglesias *et al.*, 1993; Ballicora *et al.*, 1995). The large subunit and small subunits could be expressed separately as well as together. As seen in Table II, considerable activity of ADPGlc PPase is obtained when the cDNA of the large subunit is expressed along with the cDNA of the small subunit enzyme in an *E. coli* mutant devoid of ADPGlc PPase activity. The purified recombinant enzyme, containing both the large and small subunits, has a specific activity of 64 μmol · min^{-1} · mg^{-1} when measured in the presence of the activator (3 m*M* 3PGA). If the large subunit is expressed alone, little activity is observed. However, expression of the small subunit alone leads to significant ADPGlc PPase activity (Ballicora *et al.*, 1995). This homomeric (four small subunits) enzyme has been puri-

TABLE II

COMPARISON OF THE PROPERTIES OF TRANSGENIC ADPGlc PPASES WITH THE PROPERTIES OF THE POTATO TUBER ENZYME[a]

Enzyme source	$A_{0.5}$ (m*M*)	$I_{0.5}$ (m*M*) at 0.25 m*M*, 3PGA	$I_{0.5}$ (m*M*) at 3.0 m*M*, 3PGA
Potato tuber (Sowokinos and Preiss, 1982)	0.40	0.12	0.33
pMLaugh10 + pMON17336 (large and small subunits)	0.16	0.07	0.63
pMLaugh10 (small subunit only)	2.40	—	0.08

[a] The kinetic constants of the recombinant enzyme purified from *E. coli* were measured (Ballicora *et al.*, 1995) and they coincided with the data obtained with the native potato tuber enzyme (Sowokinos and Preiss, 1982). $A_{0.5}$ and $I_{0.5}$ are concentration of activator PGA needed for 50% of maximal activation and concentration of inhibitor P_i giving 50% inhibition, respectively.

fied almost to homogeneity with a specific activity of 50 μmol · min^{-1} · mg^{-1} when measured in the presence of a high concentration (4 mM) of 3PGA. As shown in Table II, the enzyme composed exclusively of small subunits has a lower apparent affinity ($A_{0.5}$ = 2.4 mM) for the activator, 3PGA, than the heterotetramer. The enzyme with only the small subunit is also more sensitive to P_i inhibition ($I_{0.5}$ of 0.08 mM in the presence of 3 mM 3PGA) as compared with the heteromeric enzyme ($I_{0.5}$ value of 0.63 mM). The K_m values for the substrates and Mg^{+2} are essentially the same whether the enzyme is composed of only one subunit, the small subunit, or two subunits, small and large. In every case the native enzyme is a tetramer—a homotetramer in the case of the small subunit alone and a heterotetramer in the case of the large and small subunits (Ballicora *et al.,* 1995).

These data suggest that the small subunit is primarily involved in catalysis; it has substantial activity in the absence of the large subunit if the concentration of 3PGA, the activator, is high. The large subunit, when expressed alone, has little activity, but if expressed with the small subunit, the resulting enzyme has similar regulatory kinetic constants as does the native potato enzyme. This suggests that the prime function of the large subunit would be to regulate the activity of the small subunit, increasing the apparent affinity for the activator, and decreasing the affinity for the inhibitor P_i. This information agrees with results obtained with *A. thaliana,* in which the mutant ADPGlc PPase lacking the large subunit had activity but its affinity for the activator, 3PGA, was lower and the affinity for P_i was higher than for the wild-type heterotetrameric enzyme (Li and Preiss, 1992).

The small subunit of the higher plant ADPGlc PPases is highly conserved (85–95% identity), whereas the large subunit is less conserved (50–60% identity; Smith-White and Preiss, 1992). The higher heterogeneity seen in the large subunit sequence probably reflects different demands in the modulation of the small subunit sensitivity to allosteric activation and inhibition posed by different demands of the tissue and species. Expression of large subunits would differ during development or in different plants and tissues (e.g., leaf, stem, guard cells, tuber, endosperm, root, embryo), providing the resulting ADPGlc PPases with differing sensitivities to regulators.

VI. IDENTIFICATION OF THE SUBSTRATE BINDING SITES

Chemical modification can be used to obtain information on the catalytic mechanism and on the catalytic site of the enzyme of interest. One goal in the design of affinity labels for enzymes is to determine the catalytically important residues. First, the affinity label has to behave as an analogue of the substrate (or of the activator or inhibitor) by competition experiments.

Second, the enzyme is covalently bound to the affinity label in conditions chosen according to the enzyme in question and the chemical nature of the analogue, so as to decrease nonspecific labeling. Third, the labeled enzyme is subjected to proteolysis and the radioactive peptide(s) are isolated by HPLC. The labeled peptide(s) are then sequenced, providing information about the domains of the enzyme involved in the interaction with the substrate (or with the modulators).

Chemical modification studies on ADPGlc PPase have involved the use of the following affinity labels:

1. Pyridoxal-5-phosphate (PLP), an analog of 3PGA or phosphorylated sugars that can be covalently bound to the enzyme by reduction with $NaBH_4$
2. The photoaffinity substrate analogs, 8-azido-ATP and 8-azido-ADPglucose. When ultraviolet (UV) light (257 nm) irradiates 8-azido compounds, a nitrene radical is formed, which can react with electron-rich residues and inactivate the enzyme.
3. Phenylglyoxal, for the identification of arginine residues

These studies have provided information on the catalytic and regulatory sites of the spinach and cyanobacterial ADPGlc PPases, and on the role of the large and small subunits (Morell *et al.,* 1988; Smith-White and Preiss, 1992; Ball and Preiss, 1994; Charng *et al.,* 1994).

In addition, residues that chemical modification suggested were involved in substrate binding have been subjected to site-directed mutagenesis (Kumar *et al.,* 1989; Hill *et al.,* 1991; Charng *et al.,* 1994, 1995; Sheng *et al.,* 1996). These studies have provided information on the catalytic and regulatory sites of the spinach ADPGlc PPase and on the role of the large and small subunits. They have also shown that many of the studies initiated with the bacterial ADPGlc PPases are highly relevant for studies on the higher plant enzyme (Kumar *et al.,* 1988; Hill *et al.,* 1991; Charng *et al.,* 1994; Sheng *et al.,* 1996).

In the ADPGlc PPase from *E. coli,* the Lys residue 195 has been identified as the binding site for the phosphate of glucose-1-P (Hill *et al.,* 1991), and tyrosine (Tyr) residue 114 has been identified as involved in the binding of the adenosine portion of the other substrate, ATP (Lee and Preiss, 1986). When the amino acid sequence of the *E. coli* enzyme is aligned with those from the plant and cyanobacterial ADPGlc PPases, the identity ranges from 30 to 33% (Smith-White and Preiss, 1992). Sequence identity is much higher when only the ATP and glucose-1-P binding sites (Table III) are compared with the corresponding sequences of the plant and cyanobacterial enzymes, suggesting that those sequences are still important in the plant enzyme, probably having the same function.

TABLE III

CONSERVATION OF THE SEQUENCE OF THE *E. coli* ADPGlc PPASE BINDING SITES FOR Glc-1-P[a] AND ATP[b] IN THE ENZYMES FROM OTHER ORGANISMS[c]

Organism	Glc-1-P site	ATP site
Prokaryotes		
E. coli	IIEFVEKP-AN	WYRGTADAV
S. typhimurium	**D*****_**	*********
Anabaena	V*D*S***KGE	*FQ******
Synechocystis	*TD*S***QGE	*FQ******
Plant small subunit		
Spinach leaf, 51 kDa	****A***KGE	*FQ******
Potato tuber, 50 kDa	****A***QGE	*FQ******
Maize endosperm, 54 kDa	****A***KGE	*FQ******
Rice seed	*V**A***KGE	*FQ******
A thaliana	****A***KGE	*FQ******
Wheat endosperm	****A***KGE	*FQ******
Plant large subunit		
Spinach leaf, 54 kDa	VLS*S***KGD	*FQ******
Potato tuber, 51 kDa	VVQ*A***KGF	*FQ******
Maize endosperm, 60 kDa	VLQ*F***KGA	*FQ****SI
A. thaliana	V*SFS***KGD	*FQ*****L
Wheat endosperm	VVQ*S*Q*KGD	*FR*****W

[a] Data from Hill *et al.* (1991).

[b] Data from Kumar *et al.* (1988).

[c] For references to sequences, see Smith-White and Preiss (1992) for the plant enzymes; Charng *et al.* (1992) for *Anabaena;* Kakefuda *et al.* (1992) for *Synechocystis;* and Ainsworth *et al.* (1993) for the wheat endosperm small subunit. Lys-195 and Tyr-114 of the *E. coli* enzyme belong to the Glc-1-P and ATP binding sites, respectively. * signifies the same amino acid as in the *E. coli* enzyme.

The binding site for pyridoxal phosphate in the small subunit was isolated, revealing a lysine (Lys) residue close to the C terminus, which may be important for 3PGA activation (Morell *et al.,* 1988). When PLP is covalently bound (Fig. 2), the plant ADPGlc PPase no longer requires 3PGA for activation; and the binding of PLP is prevented by the allosteric effectors, 3PGA and P_i. These observations indicate that the activator analog, PLP, is binding at the activator site. In addition, Preiss *et al.* (1992) and Ball and Preiss (1994) showed that three Lys residues of the spinach leaf large subunit are also involved or are close to the binding site of pyridoxal-P and, presumably, to the activator, 3PGA (Table IV). The chemical modification of these Lys residues by pyridoxal-P was prevented by the presence

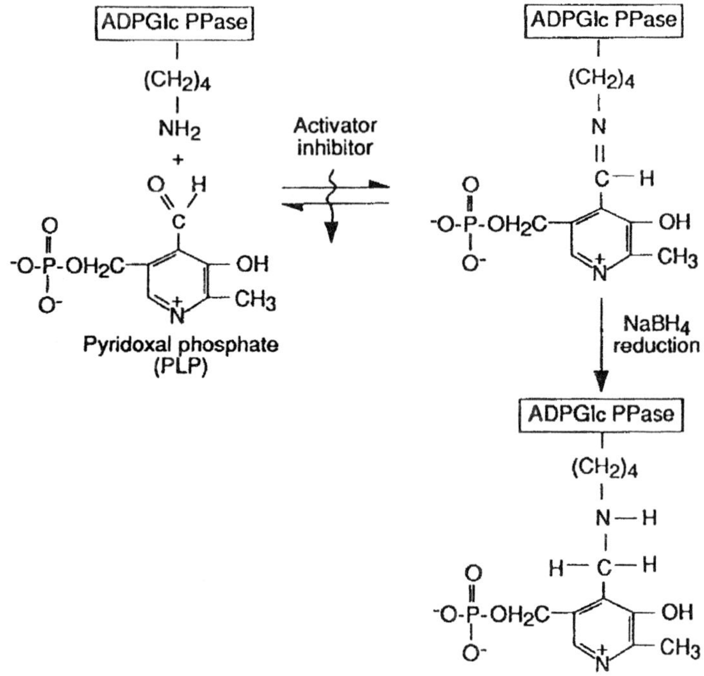

FIG. 2. Chemical modification is one of the tools used to identify the amino acid residues involved in the binding of a substrate, activator, or inhibitor. In the case of the ADPglucose pyrophosphorylase (ADPGlc PPase), the allosteric sites can be modified using pyridoxal-5-phosphate (PLP). PLP forms a Schiff base with an ε-amino group of a Lys residue. This Schiff base is converted to a stable secondary amine by reduction with $NaBH_4$. The modified enzyme no longer requires activator for catalysis, indicating that a Lys residue participates in the binding of the activator. This evidence is supported by the fact that modification of the enzyme with PLP can be prevented if an allosteric effector (i.e., 3PGA or P_i) is present when the enzyme is incubated with PLP.

of 3PGA during the reductive pyridoxylation process and, in the case of the Lys residue of site 1 of the small subunit and site 2 of the large subunit, P_i also prevented them from being modified by reductive pyridoxylation. Thus, it is believed that the most important sites involved are sites 1 and 2. Similar results were obtained with the *Anabaena* ADPGlc PPase (Charng *et al.,* 1994). Chemical modification of the enzyme with PLP caused the cyanobacterial enzyme no longer to require activator for maximal activity; chemical modification was prevented by 3PGA and P_i. The modified Lys residue was identified as Lys-419 and the sequence adjacent to that residue is similar to that observed for site 1 sequences in the higher plants. Site-directed mutagenesis of Lys-419 to either Arginine (Arg), Alanine (Ala), Glutamine (Gln), or glutamic acid (Glu) produced mutant enzymes (ex-

TABLE IV
PLANT AND CYANOBACTERIAL ADPGlc PPASE ACTIVATOR BINDING SITES[a]

	Activator site 1	Activator site 2
Potato tuber, 50 kDa	SGIVTVIKDALIPSGIII	IKRAIIDKNAR
Spinach, 51 kDa (small)	SGIVTVIKDALIPSGTVI	IKRAIIDKNAR
Maize, 54 kDa	GGIVTVIKDALLPSGTVI	IRRAIIDKNAR
Wheat seed (small)	SGIVTVIKDALLPSGTVI	IKRAIIDKNAR
Anabaena	SGIVVVLKNAVITDGTII	QRRAIIDKNAR
Synechocystis	NGIVVVIKNVTIADGTVI	IRRAIIDKNAR
Spinach, 54 kDa (large)	SGITVIFKQATIKDGVV	IKDAIIDKNAR
Potato, 51 kDa (large)	SGIIIILEKATIRDGTVI	IRKCIIDKNAK
Maize, 60 kDa (large)	SGIVVILKNATINECLVI	IRNCIIDMNAR
Wheat seed (large)	SGIVVIQKNATIKDGTVV	IQNCIIDKNAR
Barley endosperm (large)	SGIVVIQKNATIKDGTVV	ISNCIIDMNAR

[a] The sequences listed in one-letter code are from Smith-White and Preiss (1992). The sequences of the barley endosperm enzyme are from Villand *et al.* (1992). The Lys residues underlined indicate they are covalently modified by pyridoxal-P and the chemical modification of the Lys residue is prevented by 3PGA and P_i, or site-directed mutagenesis has identified them to be involved in binding the activator. The numbers 441 and 417 correspond to the Lys residues in the potato tuber ADPGlc PPase small subunit. Site 1 is present both in the large and in the small subunits of the plant ADPGlc PPase, whereas site 2 is only in the large subunit even though similar sites are observed in the small subunit.

pressed in *E. coli*) with lowered affinities, 25- to 150-fold lower than that of the wild-type enzyme. No other kinetic constants, such as affinity for substrates and the inhibitor, P_i, were affected, nor was the heat stability or the catalytic efficiency of the enzyme affected. These mutant enzymes, however, were still activated to a great extent at higher concentrations of 3PGA, suggesting that an additional site was involved in the binding of the activator. The Lys-419 in the Arg mutant was chemically modified with the activator analog, PLP, and Lys 382 was the amino acid that was reductively phosphopyridoxylated. Modification of Lys-382 in the Arg mutant also caused a dramatic alteration in the allosteric properties of the enzyme, which could be prevented by the presence of 3PGA or P_i during the chemical modification process. Therefore, Lys-382 was identified as the additional site involved in the binding of the activator and, as seen in Table IV, the adjacent sequence about Lys-382 in the *Anabaena* enzyme is similar to that seen for site 2.

In the ADPGlc PPases of *Anabaena* and higher plants, there are five highly conserved Arg residues that are not present in the enteric bacterial ADPGlc PPases. As discussed previously, the regulatory characteristic of enteric bacteria are different from those of cyanobacteria and higher plants;

for example, the enteric ADPGlc PPases are not inhibited by P_i, but by 5-AMP. Phenylglyoxal inactivation of the spinach enzyme can be prevented by 3PGA or by P_i, which is evidence that one or more Arg residues are present in the allosteric sites of the spinach leaf enzyme. Both subunits of the spinach leaf enzyme were labeled when [^{14}C]phenylglyoxal was used (Ball and Preiss, 1992). Thus, Arg residues may also be involved in the binding of the allosteric ligands, particularly P_i. Site-directed mutagenesis was used to find out whether these five Arg residues were in some way responsible for the different regulatory properties. All five conserved Arg residues in the *Anabaena* ADPGlc PPase—that is, Arg 66, 105, 171, 294, and 385 were mutagenized to Ala (Sheng and Preiss, 1998). As shown in Table V, the Arg 294 Ala mutation resulted in a mutant enzyme with a much lower affinity for the inhibitor, phosphate, measured in the absence or presence of 3PGA. This mutation had no (or little) effect on the kinetic constants for the substrates or for the activator, 3PGA (Sheng and Preiss, 1998), and it can be concluded that Arg 294 of the *Anabaena* enzyme is involved in the binding of P_i. The activator, 3PGA, and the inhibitor, P_i, probably bind to different sites, although there could be some overlapping.

Another effect of the site-directed mutagenesis was that the purified mutant enzyme Arg 294 Ala had a 3-fold higher specific activity than the wild-type enzyme, suggesting that with disappearance of the inhibitor binding site there was also a conformational change, resulting in an enzyme with a higher catalytic efficiency. These results not only clarified another aspect of the structure–function relationships of the ADPGlc PPase, but also resulted in the creation of an enzyme that might be useful in the development of transgenic crops with higher starch production.

TABLE V

EFFECT OF SITE-DIRECTED MUTAGENESIS OF SEVERAL AMINO ACIDS ON THE RESPONSE OF THE *Anabaena* ADPGlc PPASE TO 3-PGA

	3-P-glycerate	WT	R66A	R105A	R294A	R385A
$I_{0.5}$ P_i (m*M*)	—	0.055	0.26	0.077	5.2	0.062
	+	1.0	0.58	0.89	38	0.87
V_{max} (unit[a]/mg)	—	6.9	4.8	4.8	11	0.63
	+	60	44	79	170	13

[a] One unit of enzyme activity is defined as the amount of enzyme required to form 1 μmol of ADP-glucose/min at 37°C (assay in the direction of synthesis).

As discussed previously, cDNA clones encoding the putative mature forms of the large and small subunits of the potato tuber ADPGlc PPase have been expressed together, using two different compatible vectors, in an *E. coli* mutant deficient in ADPGlc PPase activity (Iglesias *et al.*, 1993; Ballicora *et al.*, 1995; Table II). This expression system was then used for site-directed mutagenesis experiments aiming to test whether the Lys residues in the potato tuber ADPGlc PPase have a role in activation, as suggested by the chemical modification (with pyridoxal-P) experiments of the spinach enzyme.

As shown in Table VI, site-directed mutagenesis of Lys 441 of the potato ADPGlc PPase small subunit to Glu and Ala results in mutant enzymes with lower affinity, 30- to 83-fold, respectively, for 3PGA (Ballicora *et al.*, 1996; Preiss *et al.*, 1996). A conservative mutation to arginine resulted in only a two-fold increase in $A_{0.5}$, indicating that the positive charge of the cationic amino acid is important for the binding of the activator. Mutagenesis of Lys residue 417 in the large subunit (the residue homologous to the *Anabaena* Lys residue 382 and to site 2 of the spinach leaf large subunit Lys residue modified by PLP) was also done. When Lys 417 was replaced by either Ala or Glu, the affinity for 3PGA decreased (Table V) but the increase in $A_{0.5}$ was only 3- to 13-fold and not as high as seen with the mutations of the small, 50-kDa subunit Lys 441 residue. When both Lys residues in the large (51-kDa) and small subunits were mutated, the decrease in affinity or increase in $A_{0.5}$ was additive. Thus, Lys residues in both subunits seem to contribute to the binding of the activator.

TABLE VI

SITE-DIRECTED MUTAGENESIS OF LYS RESIDUES AT THE BINDING SITE FOR THE ALLOSTERIC ACTIVATOR IN THE SUBUNITS OF THE POTATO TUBER ADPGlc PPASE. EFFECT ON THE SENSITIVITY OF THE HOLOENZYME TO THE ACTIVATOR, 3PGA[a]

ADPGlc PPase subunits			
Large	Small	3PGA $A_{0.5}$ (m*M*)	Ratio of $A_{0.5}$ mutant/$A_{0.5}$ wt
Wild-type	Wild-type	0.10	1
K417A	Wild-type	0.3	3
K417E	Wild-type	1.3	13
K417A	K441A	6.0	60
K417E	K417E	No activation	0
Wild-type	K441R	0.18	1.8
Wild-type	K441A	3.2	32
Wild-type	K441E	8.3	83

[a] Data from Preiss *et al.* (1996) and unpublished results of M. A. Ballicora and J. Preiss.

Random mutagenesis has also been used to determine whether other sequence regions or amino acids in the large subunit are important for the allosteric function (Greene *et al.,* 1996a,b). In one study (Greene *et al.,* 1996b), the Asp residue 416 (413 in the special notation used by Greene *et al.*) was mutated to an Ala residue and the affinity for 3PGA decreased about 6-fold, similar to the decrease observed when Lys 417 was mutated to Ala (Ballicora *et al.,* 1996; Table IV). In a second mutant isolated via random mutagenesis, Leu had replaced the proline residue 52 (Greene *et al.,* 1996a). The mutant enzyme's affinity for 3PGA was substantially decreased; the $A_{0.5}$ being increased 45-fold in mutant P52L, suggesting that a region of the large subunit *N*-terminal may also be involved in the formation of the allosteric activator binding site.

Giroux *et al.* (1996) described the effect of a single gene mutation in the *sh* 2 locus of maize (coding for the large subunit of the ADPGlc PPase), which increases seed weight by 11 to 18% without changing the proportion of the seed weight taken by starch. The direct effect of the mutation is the addition of two amino acids, tyrosine and serine, that seem to decrease the sensitivity of the ADPGlc PPase to inhibition by phosphate. This change in regulatory properties was found in the ADPGlc PPase measured in the seed extract and in the enzyme expressed in *E. coli.* When the researchers placed the two extra amino acids in the corresponding position of the potato tuber ADPGlc PPase, expressed in *E. coli,* they observed a similar decrease in sensitivity to P_i.

VII. CLONING OF THE ADPGlc PPase GENES AND COMPARISON OF THEIR SEQUENCES

Many cDNA or genomic clones for the small subunit ADPGlc PPase gene of rice endosperm (Krishnan *et al.,* 1986; Anderson *et al.,* 1989, 1990), maize endosperm (Barton *et al.,* 1986), spinach leaf (Preiss *et al.,* 1989), *A. thaliana* (B. Smith-White and J. Preiss, unpublished results, 1998), and potato tuber (Anderson *et al.,* 1990; Nakata *et al.,* 1991) have been isolated. In addition, a cDNA clone for the maize endosperm ADPGlc PPase large molecular subunit (*Sh* 2 locus) has also been isolated (Barton *et al.,* 1986). Olive *et al.* (1989) isolated cDNA clones from wheat leaf and wheat endosperm, which are now considered to represent the large subunit gene of the ADPGlc PPase, as suggested by the deduced amino acid sequence.

Although the isolation of the spinach leaf large subunit cDNA clone has not been reported, the major portion of the spinach leaf large subunit (54 kDa) has been sequenced by the Edmann degradation technique (B. Smith-White and J. Preiss, 1992). Since 1991, many other ADPGlc PPase genes, either genomic or represented by a cDNA, have been isolated from many plants and different tissues, and they are too numerous to cite here. Figure 3 shows the deduced amino acid sequences of 45 subunits of ADPGlc PPases obtained from the EMBL nucleotide sequence library and GenBank.

At the DNA level, the isolated genes are dissimilar. For example, in wheat leaf and wheat endosperm, there is only 55.7% identity (Olive *et al.*, 1989) and, on the basis of Southern blot hybridization analyses and restriction enzyme mapping, it is concluded that there are at least two distinct gene families in wheat. For spinach leaf and rice endosperm, there is only approximately a 50% identity (B. S. White and J. Preiss, unpublished results, 1998).

Good identity is observed in comparing amino acid sequences of similar subunits of the ADPGlc PPase from the different plants, and this is expected since the spinach leaf lower-molecular-weight subunit antibody reacts well with the equivalent subunits of maize endosperm (Plaxton and Preiss, 1987; Preiss *et al.*, 1990), rice seed (Krishnan *et al.*, 1986; Anderson *et al.*, 1989), *Arabidopsis* leaf (Lin *et al.*, 1988a,b), and potato tuber (Okita *et al.*, 1990) enzymes. The lower-molecular-weight antibody does not react well with the higher-molecular-mass subunit of the ADPGlc PPase of these various plants. Therefore, it was not expected that much homology would be seen between the lower- and higher-molecular-weight subunits. However, there appears to be some identity (approximately 40–60%) between the large and small subunits of the higher plant ADPGlc PPase (Fig. 3).

Because of the relatively low but certain homology between the two subunits of the ADPGlc PPase, it can be speculated that they may have arisen originally from the same gene. The bacterial ADPGlc PPase is a homotetramer composed of only one subunit (Preiss, 1984). The cyanobacterial ADPGlc PPase has 3PGA as an allosteric activator and P_i as an inhibitor, similar to the enzyme from higher plants (Levi and Preiss, 1976), and unlike the bacterial enzymes (e.g., fructose-1,6-biphosphate is the activator in enteric bacteria). Both bacterial (Preiss, 1984; Preiss and Romeo, 1989) and cyanobacterial (Iglesias *et al.*, 1991) ADPGlc PPases are homotetrameric, unlike the higher plant enzymes, indicating that regulation by 3PGA and P_i (a good signaling system for a photosynthetic organism) is not related to the heterotetrameric nature of the higher plant enzyme. It is possible that during evolution there was duplication of the ADPGlc PPase gene, and divergence of the genes then produced two different genes coding for the two peptides, both of which were required for optimal activity of the native higher plant enzyme.

As indicated in the preceding, one can tentatively assign catalytic function to the small subunit of the ADPGlc PPase. The extensive identity and

a

Name: d21272:	cDNA, RT PCR, *Oryza sativa*, callus
Name: x91736:	cDNA, *Chlamydomonas reinhardti*
Name: x76941:	cDNA, *Vicia faba*, var. minor cv. Fribo, cotyledons
Name: x96764:	cDNA, Pisum sativum, cv. sugar snap, cotyledons
Name: u11281:	cDNA, *Ipomoea batatas*, strain White Star
Name: x83498:	cDNA, *Ipomoea batatas*
Name: z46756:	cDNA, *Ipomoea batatas*, strain White Star
Name: x76940:	cDNA, *Vicia faba*, var. minor cv. Fribo, cotyledons
Name: x96765:	cDNA, Pisum sativum, cv. sugar snap, cotyledons
Name: x83500:	cDNA, *Spinacia oleracea*
Name: l33648:	genomic, *Solanum tuberosum*, cv. Russett Burbank
Name: x61186:	cDNA, *Solanum tuberosum*, cv. Russett Burbank, tuber
Name: x55155:	cDNA, *Solanum tuberosum*, cv. Desiree {3112 Ebstorf}, tuber
Name: x55650:	cDNA, *Solanum tuberosum*, cv. Desiree {3112 Ebstorf}, tuber
Name: l41126:	cDNA, *Lycopersicon esculentum*, fruit
Name: x78899:	cDNA, *Beta vulgaris*, cv. Zuchtlinie 5S0026, tap root
Name: j04960:	cDNA, *Oryza sativa*, strain L.C.V. Biggs M201, endosperm
Name: m31616:	cDNA, *Oryza sativa*, strain L.C.V. Biggs M201, leaf
Name: x62241:	cDNA, RT PCR, *Hordeum vulgaris*, cv Bomi, endosperm (S39537)
Name: x66080:	cDNA, *Triticum aestivum*, cv. Chinese Spring, leaf
Name: z48562:	cDNA, *Hordeum vulgaris*, cv Bomi, starchy endosperm
Name: z48563:	cDNA, *Hordeum vulgaris*, cv Bomi, leaf
Name: x73365:	cDNA, RT PCR, *Arabidopsis thaliana*
Name: brittle2:	cDNA, *Zea mays*, endosperm, *brittle-2* locus
Name: s72425:	cDNA, *Zea mays*, leaf
Name: atsmall:	cDNA, *Arabidopsis thaliana*, above-ground (B. Smith-White, pers. comm.)
Name: atlarge:	cDNA, *Arabidopsis thaliana*, above-ground (B. Smith-White, pers. comm.)
Name: x73367:	cDNA, RT PCR, *Arabidopsis thaliana*
Name: x14348:	cDNA, *Triticum aestivum*, cv. Mardler, leaf

Name: x62243:	cDNA, RT PCR, *Hordeum vulgaris*, cv Bomi, 4 week seedling, (S39540)
Name: x76136:	cDNA, *Solanum tuberosum*, cv. Desiree {3112 Ebstorf}, tuber
Name: pcrcode:	genomic PCR, Spinacia oleracea (B. Smith-White, pers. comm.)
Name: x96766:	cDNA, Pisum sativum, cv. sugar snap, cotyledons
Name: x61187:	cDNA, *Solanum tuberosum*, cv. Russett Burbank, tuber
Name: x73366:	cDNA, RT PCR, *Arabidopsis thaliana*
Name: x78900:	cDNA, *Beta vulgaris*, cv. Zuchtlinie 5S0026, tap root
Name: x74982:	cDNA, *Solanum tuberosum*, cv. Desiree {3112 Ebstorf}, leaf
Name: x73364:	cDNA, RT PCR, *Arabidopsis thaliana*
Name: x14349:	cDNA, *Triticum aestivum*, cv. Mardler, endosperm
Name: z21969:	cDNA, *Triticum aestivum*, cv. Chinese Spring, developing grain
Name: x62242:	cDNA, RT PCR, *Hordeum vulgaris*, cv Bomi, endosperm (S39540)
Name: x67151:	cDNA, *Hordeum vulgaris*, cv Bomi, endosperm
Name: x14350:	cDNA, *Triticum aestivum*, cv. Mardler, endosperm
Name: z38111:	cDNA, *Zea mays*, embryo
Name: s48563:	cDNA, *Zea mays*, endosperm, *shrunken-2* locus

b

Five consensus classes:

ONLYSMLL - residue found only in small subunit class, diagnostic for class membership
ALLSMALL - residue found in all members of small subunit class,
ONLYLRGE - residue found only in large subunit class, diagnostic for class membership
ALLLARGE - residue found in all members of large subunit class,
ALLPLANT - residue found in all plant proteins.

Uppercase - no exceptions, lowercase - one or two exceptions

FIG. 3.

```
C         1                                                      51                                               100
ALLSMALL                                                                                                              S LGI
ONLYSMLL                                                                                                              S LGI
 d21272   .......... .......... .......... .......... .......... .......... .......... .......... .......... ..........
 x91736   .......... ...MALAVRP AVASRAAGKV NLRSGSAAAC RVVRLSVSAE PATKAATFSN QISRPSQTVD RVSSYIALNT SQLASDSKNV SALKSVLGII 87
 x76941   ........MA AIGVLKVPP. .SSSSSSSSS SSKAIARNLS FTSSHLSGDK IFTLSGR... .TRRTSGRNP FIVSPKAVSD SKNSQTCLDP DASRSVLGII 86
 x96764   .....MASMA AIGVLKVPPS SSSSSSSSSS SSKAIARNLS FTSSQLCGDK IFTVSG.... .TRRSSGRNP FIVSPKAVSD SKNSQTCLDP DASRSVLGII 90
 u11281   .......... .......... .......... .......... .......... .......... .......... .......... .......... ..........
 x83498   .......... .......... .......... .......... .......... .......... .......... .......... .......... ..........
 z46756   .......... .......... .......... .......... .......... .......... .......... .......... .......... ..........
 x76940   .......... ......MSSI VTSGVINVPR SSSSSKNLSF SSSSQLSGNK ILTVSG..NG APRGRCTLKH VFLTPKAVSD SQNSQTCLDP DASRSVLGII 82
 x96765   .......... ......MSSI VTSSVINVPR SSSSSKNLSF .SSSQLSGDK ILTVSG..KG APRGRCTRKH VIVTPKAVSD SQNSQTCLDP DASRSVLGII 81
 x83500   .......... .......... .......... .......... .......... .......... .......... .......... ..NSQTCLDP EASRSVLGII 18
 l33648   .....MAASI GALKSSPSSN NCINERRNDS TRAVSSRNLS FSSSHLAGDK LMPVSSLRSQ GVRFNVRRSP MIVSPKAVSD SQNSQTCLDP DASRSVLGII 95
 x61186   .....MAASI GALKSSPSSN NCINERRNDS TRAVSSRNLS FSSSHLAGDK LMPVSSLRSQ GVRFNVRRSP MIVSPKAVSD SQNSQTCLDP DASRSVLGII 95
 x55155   .......... .......... .......... .......... .......... .......... .......... .......... ....QTCLDP DASRSVLGII 16
 x55650   .......... .......... .......... .......... .......... .......... .......... .......... ....QTCLDP DASRSVLGII 16
 l41126   .......... .......... .......... .......... .......... .......... .......... .......... .......... MLAGVFWVII 10
 x78899   .......... .......... ....ITVPST SSKNLQNSLA FSSSSLSGDK IQTTSFLNRR YCRISS.RAP IVVSPKAVSD SKNSQTCLDP EASRSVLGII 75
 j04960   .......... .......... .......... .......... ......MNVL ASKIFPSRSN VVSEQQQSKR E.KATIDDAK NSSKNKNLDR SVDESVLGII 53
 m31616   .......... .......... .......... .......... ......MNVL ASKIFPSRSN VASEQQQSKR E.KATIDDAK NSSKNKNLDR SVDESVLGII 53
 x62241   .......... .......... .......... .......... .......... .......... .......... .......... .......... ..........
 x66080   .......... .......... .......... .......... .......... ..MDVPLASK TFPSPSPSKR E.QCNIDGHK SSSKHADLNP HVDDSVLGII 47
 z48562   .......... .......... .......... .......... .......... ..MDVPLASK .VPLPSPSKH E.QCNVYSHK SSSKHADLNP HAIDSVLGII 46
 z48563   .......... ...MAMAAAA SPSKILIPPH RASAVTAAAS TSCDSLRLLC APRGRPGPRG LVARPVPRRP FFFSPRAVSD SKSSQTCLDP DASTSVLGII 87
 x73365   .......... .......... .......... .......... .......... .......... .......... .......... .......... ..........
brittle2  .......... .......... .......... .......... .......... .MDMALASKA SPPPWNATAA EQPIPKRDKA AARDSTYLNP QAHDSVLGII 49
 s72425   .......... .......... .......... .......... .......... .......... .......... .......... .......... ..........
 atsmall  .......... .......... .......... .......... .........D KISLKSTVSR LCKSVVRRNP IIVSPKAVSD SQNSQTCLDP DASSSVLGII 51
ALLPLANT                                                                                                        V  i
 atlarge  .......... .......... .......... .......... ....MGKKLN LSQLPNIRLR SSTNFSQKRI LMSLNSVAGE SKVQELETEK RDPRTVASII 56
 x73367   .......... .......... .......... .......... .......... .......... .......... .......... .......... ..........
 x14348   .......... .......... .......... .......... .......... .......... .......... .......... .......... ..........
 x62243   .......... .......... .......... .......... .......... .......... .......... .......... .......... ..........
 x76136   .......... .......... .......... .......... .......MGK KLKYTKFQLR SNVVKPNICM SLTTDIAGEA KLKDLERQKK GDARTVVAII 53
 pcrcode  .......... .......... .......... .......... .......... .......... .......... .......... .......... ..........
 x96766   .......... ........MA SGCVSLKTNT HFPNSKKGSF FGERIKGSLK NSSWVTTQKK IKPASFSAIL TSDDPKGSLN LQVPSFLRLR ADPKNVISIV 82
 x61187   .......... .......... .......... .......... .......... ........NK IKPGVAYSVI TTENDTQTVF VDMPRLERRR ANPKDVAAVI 42
 x73366   .......... .......... .......... .......... .......... .......... .......... .......... .......... ..........
 x78900   ....MDASAA AINVNAHLTE VGKKR..FLG ERISQSLKGK DLRALFSRTE SKGRNVNKPG VAFSVLTSDF NQSVKESLKY EPALFES.PK ADPKNVAAIV 93
 x74982   ....MDALCA SMKGTAQLVA ICNQESAFWG EKISGRRLIN KGFGVRS.CK SFTTQQRGRN VTPAVLTRDI N...KEMLPF EESMFEEQPT ADPKAVASVI 92
 x73364   .......... .......... .......... .......... .......... .......... .......... .......... .......... ..........
 x14349   .......... .......... .......... .......... .......... .......... .......... .......... .......... ..........
 z21969   MSSMQFSSVL PLEGKACISP VRREGSASER LKVG.DSSSI RHERASRRMC NGGRGPAATG AQCVLTSDAS PADTLVLRTS FRRNYA.... .DPNEVAAVI 95
 x62242   .......... .......... .......... .......... .......... .......... .......... .......... .......... ..........
 x67151   MSSMQFSSVL PLEGKACVSP VRREGSACER LKIG.DSSSI RHERASRRMC NGGAGAPPPP VRSACSPPTP ARPTPLFSGR PSGGITPIRT RSRPSVAAVI 99
 x14350   .......... .........R ASPPSESRAP LRAP.QRSAT RQHQARQGPR RMCNGGRGPP YWTAGVTSAP ARQTPLFSGR PSGGLS.... .DPNEVAAVI 75
 z38111   ...MQFSSVL PLEGKACMSP VRRGSGGYGS ERMRINCCSI RRNKALRRMC FSARGAVSST .QCVLTSDAG PDTLVRPNHP FRRNYA.... .DPNEVAAVI 91
 s48563   .....MQFAL ALDTNSGPHQ IRSCEGDGID RLEKLSIGGR KQEKALRNRC FGGRVAA..T QQCILTSDCA PETLHSQTQS SRKNYA.... .DNARVSAII 88
ALLLARGE                                                                                                      d     a
ONLYLRGE                                                                                                            a
```

d

```
           101                                                          151                                                        200
ALLSMALL       A  R Y    KK  K     L AN     I V   L   N S IYV       a     L S A   N  G  YKNE F       Q S  Np ..
ONLYSMLL                  K        L A        V   L   N S                          G  YKNE          Q S  Np .
d21272     .......... .......... .......... .......... .......... .......... .RTYNIGEGV GF.GDGFVEV LAATQTTGES GKRWFQGTAD 38
x91736     EFKRTGTRLF PLTKSRAKPA VPIGGAYRLI DVPMSNCINS GISKIYILTQ FNSTSLNRHL GRAYNMGSGV RFGGDGFVEV LAATQTPTD. .KEWFQGTAD 185
x76941     LGGGAGTRLY PLTKKRAKPA VPLGANYRLI DIPVSNCLNS NISKIYVLTQ FNSASLNRHL SRAYASNLGG .YKNEGFVEV LAAQQSPENP N..WFQGTAD 183
x96764     LGGGAGTRLY PLTKKRAKPA VPLGANYRLI DIPVSNCLNS NISKIYVLTQ FNSASLNRHL SRAYASNLGG .YKNEGFVEV LAAQQSPENP N..WFQGTAD 187
u11281     .......... .......... .......... .......... .......... .......... .......... .......... .......... ..........
x83498     LGGGAGTRLY PLTKKRAKPA VPLGANYRLI DIPVSNCLNS NVSKIYVITQ FNSAYLNRHL SRAYASNMGG .YKNEGFVEV LAAQQSPENL ...WSQGTAD 96
z46756     .......... .......... .......... .......... .......... .......... .......... .......... .......... ..........
x76940     LGGGAGTRLY PLTKKRAKPA VPLGANYRLI DIPVSNCLNS NISKIYVLTQ FNSASLNRHL SRAYASNLGG .YKNEGFVEV LAAQQSPENP N..WFQGTAD 179
x96765     LGGGAGTRLY PLTKKRAKPA VPLGANYRLI DIPVSNCLNS NISKIYVLTQ FNSASLNRHL SRAYASNLGG .YKNEGFVEV LAAQQSPENP N..WFQGTAD 178
x83500     LGGGAGTRLY PLTKKRAKPA VPLGANYRLI DIPVSNCLNS NISKIYVLTQ FNSASLNRHL SRAYASNLGG .YKNEGFVEV LAAQQSPENP D..WFQGTAD 115
l33648     LGGGAGTRLY PLTKKRAKPA VPLGANYRLI DIPVSNCLNS NISKIYVLTQ FNSASLNRHL SRAYASNMGG .YKNEGFVEV LAAQQSPENP D..WFQGTAD 192
x61186     LGGGAGTRLY PLTKKRAKPA VPLGANYRLI DIPVSNCLNS NISKIYVLTQ FNSASLNRHL SRAYASNMGG .YKNEGFVEV LAAQQSPENP D..WFQGTAD 192
x55155     LGGGAGTRLY PLTKKRAKPA VPLGANYRLI DIPVSNCLNS NISKIYVLTQ FNSASLNRHL SRAYASNMGG .YKNEGFVEV LAAQQSPENP D..WFQGTAD 113
x55650     LGGGAGTRLY PLTKKRAKPA VPLGANYRLI DIPVSNCLNS NISKIYVLTQ FNSASLNRHL SRAYASNMGG .YKNEGFVEV LAAQQSPENP H..WFQGTAD 113
l41126     LGGGAGTRLY PLTKKRAKPA VPLGANYRLI DIPVSNCLNS NISKIYVLTQ FNSASLNRHL SRAYASNMGE .YKNEGFVEV LAAQQSPENP D..WFQGTAD 107
x78899     LGGGAGTRLY PLTKKRAKPA VPLGANYRLI DIPVSNCLNS NISKIYVLTQ FNSASLNRHL SRAYASNMGG .YKNEGFVEV LAAQQSPENP N..WFQGTAD 172
j04960     LGGGAGTRLY PLTKKRAKPA VPLGANYRLI DIPVSNCLNS NISKIYVLTQ FNSASPNRHL SRAYGNNIGG .YKNEGFVEV LAAQQSPDNP N..WFQGTAD 150
m31616     LGGGAGTRLY PLTKKRAKPA VPLGANYRLI DIPVSNCLNS NISKIYVLTQ FNSASLNRHL SRAYGNNIGG .YKNEGFVEV LVAQQSPDNP N..WFQGTAD 150
x62241     .......... .......... .......... .......... .......... .......... .......... .......... .......... ..........
x66080     LGGGAGTRLY PLTKKRAKPA VPLGANYRLI DIPVSNCLNS NISKIYVRTQ FNSASLNRHL SRAYGSNIGG .YKNEGFVEV LAAQQSPDNP D..WFQGTAD 144
z48562     LGGGAGTRLY PLTKKRAKPA VPLGANYRLI DIPVSNCLNS NISKIYVLTQ FNSASLNRHL SRAYGSNIGG .YKNEGFVEV LAAQQSPDNP D..WFQGTAD 143
z48563     LGGGAGTRLY PLTKKRAKPA VPLGANYRLI DIPVSNCLNS NISKIYVLTQ FNSASLNRHL SRAYGSNIGG .YKNEGFVEV LAAQQSPDNP D..WFQGTAD 184
x73365     .......... .......... .......... .......... .......... .......... .......... .......... .......... ..........
brittle2   LGGGAGTRLY PLTKKRAKPA VPLGANYRLI DIPVSNCLNS NISKIYVLTQ FNSPSLNRH. SRAYGSNIGG .YKNEGFVEV LAAQQSPDNP N..WFQGTAD 145
s72425     .......... .......... .......... .......... .......... .......... .......... .......... .......... ..........
atsmall    LGGGAGTRLY PLTKKRAKPA VPLGANYRLI DIPVSNCLNS NISKIYVLTQ FNSASLNRHL SRAYASNMGG .YKNEGFVEV LAAQQSPENP N..WFQGTAD 148
ALLPLANT   LGGG GT L  PLT  RA PA VP G  YRLI D P SNC NS  I K    tQ fNS sLNRH   R Y    G        G VEV lAA Q p      WFQGTAD
atlarge    LGGGAGTRLF PLTKRRAKPA VPIGGAYRLI DVPMSNCINS GINKVYILTQ YNSASLNRHF SRAYN.SNGL .GFGDGYVEV LAATQTPGES GKRWFQGTAD 154
x73367     .......... .......... .......... .......... .......... .......... .......... .......... .......... ..........
x14348     .......... .......... .......... .......... .......... .......... .......... .......... .......... ..........
x62243     .......... .......... .......... .......... .......... .......... .......... .......... .......... ..........
x76136     LGGGAGTRLF PLTKRRAKPA VPMGGAYRLI DVPMSNCINS GINKVYILTQ FNSASLNRHI ARAYNFGNGV .TFESGYVEV LAATQTPGEL GKRWFQGTAH 152
pcrcode    .......... .......... ......YRLI DVPISNCINS GINKVFILTQ FNSASLNRHI YRTY.HGNGI .NFGDGFVEV LAATQTQGET GKNWFQGTAD 72
x96766     LGGGPGTHLY PLTKRAATPA VPVGGCYRLI DIPMSNCINS GINKIFVLTQ FNSASLNRHI ARTY.FGNGV .NFGDGFVEV LAATQTPGEA GKKWFQGTAD 180
x61187     LGGGEGTKLF PLTSRTATPA VPVGGCYRLI DIPMSNCINS AINKIFVLTQ YNSAPLNRHI ARTY.FGNGV .SFGDGFVEV LAATQTPGEA GKKWFQGTAD 140
x73366     .......... .......... .......... .......... .......... .......... .......... .......... .......... ..........
x78900     LGGGAGTRLF PLTSRRAKPA VPIGGCYRLI DVPMSNCINS GIRKIFILTQ FNSFSLNRHL ARTYNFGDGV .NFGDGFVEV FAATQTPGES GKKWFQGTAD 193
x74982     LGGGVGTRLF PLTSRRAKPA VPIGGCYRLI DVPMSNCINS GIRKIFILTQ FNSFSLNRHL A.TYNFGNGV .GFGDGFVEV LAGTQTPGDG RKMWFQA.AD 189
x73364     .......... .......... .......... .......... .......... .......... .......... .......... .......... ..........
x14349     .......... .......... .......... .......... .......... .......... .......... .......... .......... ..........
z21969     LGGGTGTQLF PLTSTRATPA VPIGGCYRLI DIPMSNCFNS GINKIFVMTQ FNSASLNRHI HRTY.LGGGI .NFTDGSVEV LAATQMPGEA AG.WFRGTAD 192
x62242     .......... .......... .......... .......... .......... .......... .......... .......... .......... ..........
x67151     LGGGTGTQLF PLTSTRATPA VPIGGCYRLI DIPMSNCFNS GINKIFVMTQ FNSASLNRHI HRTY.LGGGI .NFTDGSVEV LAATQMPGEA AG.WFRGTAD 196
x14350     LGGGTGTQLF PLTSTRATPA VPIGGCYRLI DIPMSNCFNS GINKIFVMTQ FNSASLNRHI HRTY.LGGGI .NFTDGSVEV LAATQMPGEA AG.WFRGTAD 172
z38111     LGGGTGTQLF PLTSTRATPA VPIGGCYRLI DIPMSNCFNS GINKIFVMTQ FNSASLNRHI HRTY.LGGGI .NFTDGSVEV LAATQMPGEA AG.WFQGTAD 188
s48563     LGGGTGSQLF PLTSTRATPA VPVGGCYRLI DIPMSNCFNS GINKIFVMSQ FNSTSLNRHI HRTY.LEGGI .NFADGSVQV LAATQMPEEP AG.WFQGTAD 185
ALLLARGE            f                gc          m    g    f        a     i   t   g     F D        T   gE
ONLYLRGE            f                gc          m    g    f                      g     F D        T   gE
```

FIG. 3. *Continued*

64

e

```
          201                                                    251                                                    300
ALLSMALL  AV QYL L   E..H   V E  L  A  H      E FiQA  RE d    v AAL MDE    TAF  M   E E RI E a       eq  A      TI GLd
ONLYSMLL      YL     E.       E              E  i A    E        AA M       TA         E    E             eq
d21272    AVRQFLWLFE D.ARLKRIEN ILILSGDHLY RMDYMDFVR. SMLKGADISV ACVPV..... .......... .......... .......... .......... 91
x91736    AVAQYSWLLE D.TKNRAIED VLILSGDHLY RMDYMKFVNY HRETNADITI GCIAYGSDRA KEFGLMKIDE KRRVTSFAEK PKTQEALDAM KVDTTVLGLT 284
x76941    AVRQYLWLFE E..HN..VLE YLVLAGDHLY RMDYERFIQA HRESDADITV AALPMDEARA TAFGLMKIDE EGRIIEFSEN PKG.EQLKAM KVDTTILGLD 278
x96764    AVRQYLWLFE E..HN..VLE YLVLAGDHLY RMDYERFIQA HRESDADITV ASLPMDEARA TAFGLMKIDE EGRIVEFSEK PKG.EQLKAM KVDTTILGLD 282
u11281    .......... .......... .......... RMDYERFIQA HRETDADITV AALPMDEKRA TAFGLMKIDE EGRIIEFAEK PKR.EQLKAM KVDTTILGLD 69
x83498    AVRQYLWLFE E..HNVLVLE FLVLAGDHLY RMDYERFIQA HRETDADITV AALPMDEKRA TAFGLMKIDE EGRIIEFAEK PKR.EQLKAM KVDTTILGLD 193
z46756    .......... .......... .......... .MDYERFIQA HRETDADITV AALPMDEKRA TAFGLMKIDE EGRIIEFAEK PKR.EQLKAM KVDTTILGLD 68
x76940    AVRQYLWLFE E..HN..VLE YLILAGDHLY RMDYEKFIQA HRESDADITV AALPMDEKRA TAFGLMKIDE EGRIIEFAEK PKG.EQLKAM KVDTTILGLD 274
x96765    AVRQYLWLFE E..HN..VLE YLILAGDHLY RMDYEKFIQA HRESDADITV AALPMDEKRA TAFGLMKIDE EGRIIEFAEK PKG.EQLKAM KVDTTILGLD 273
x83500    AVRQYLWLFE E..HN..VME FLILAGDHLY RMDYERFIQA HRETDADITV AALPMDEKRA TAFGLMKIDE EGRIIEFAEK PKG.EQLQAM KVDTTILGLD 210
l33648    AVRQYLWLFE E..HT..VLE YLILAGDHLY RMDYEKFIQA HRETDADITV AALPMDEKRA TAFGLMKIDE EGRIIEFAEK PQG.EQLQAM KVDTTILGLD 287
x61186    AVRQYLWLFE E..HT..VLE YLILAGDHLY RMDYEKFIQA HRETDADITV AALPMDEKRA TAFGLMKIDE EGRIIEFAEK PQG.EQLQAM KVDTTILGLD 287
x55155    AVRQYLWLFE E..HT..VLE YLILAGDHLY RMDYEKFIQA HRETDADITV AALPMDEKRA TAFGLMKIDE EGRIIEFAEK PQG.EQLQAM KVDTTILGLD 208
x55650    AVRQYLWLFE E..HT..VLE YLILAGDHLY RMDYEKFIQA HRETDADITV AALPMDEKRA TAFGLMKIDE EGRIIEFAEK PQG.EQLQAM KVDTTILGLD 208
l41126    AVRQYLWLFE E..HN..VLE YLILAGDHLY RMDYEKFIQA HRETDADITV AALPMDEKRA TAFGLMKIHE EGRIIEFAEK PQG.QQLQAM KVDTTILGLD 202
x78899    AVRQYLWLFE E..HN..VLE YLILAGDHLY RMDYERFVQA HRETDADITV AALPMDEKRA TAFGLMKIDE EGRIIEFAEK PKG.EQLKAM KVDTTILGLD 267
j04960    AVRQYLWLFE E..HN..VME FLILAGDHLY RMDYEKFIQA HRETDSDITV AALPMDEKRA TAFGLMKIDE EGRIVEFAEK PKG.EQLKAM MVDTTILGLD 245
m31616    AVRQYLWLFE E..HN..VME FLILAGDHLY RMDYEKFIQA HRETDSDITV AALPMDEKRA TAFGLMKIDE EGRIVEFAEK PKG.EQLKAM MVDTTILGLD 245
x62241    .......... .......... .......... .......... .......... .......... .......... .......... .......... ..........
x66080    AVRQYLWLFE E..HN..VME YLILAGDHLY RMDYEKFIQA HRETDADITV AALPMDEERA TAFGLMKIDE EGRIIEFAEK PKG.EQLKAM MVDTTILGLD 239
z48562    AVRQYLWLFE E..HN..VME YLILAGDHLY RMDYEKFIQA HRETDADITV AALPMDEERA TAFGLMKIDE EGRIIEFAEK PKG.EQLKAM MVDTTILGLE 238
z48563    AVRQYLWLFE E..HN..VME YLILAGDHLY RMDYEKFIQA HRETDADITV AALPMDEERA TAFGLMKIDE EGRIIEFAEK PKG.EQLKAM MVDTTILGLE 279
x73365    .......... .......... .......... .......... .......... .......... .......... .......... .......... ..........
brittle2  AVRQYLWLFE E..HN..VME FLILAGDHLY RMDYEKFIQA HRETNADITV AALPMDEKRA TAFGLMKIDE EGRIIEFAEK PKG.EQLKAM MVDTTILGLD 240
s72425    .......... .......... .......... .......... .......... .......... .......... .......... .......... ..........
atsmall   AVR.YLWLFE E..HN..VLE YLILAGVHLY RMDYEKFIQA HRETDADITL AALPMDEQRA TAYGLMKIDE EGRIIEFAEK PKG.EHLKAM KVDTTILGLD 242
ALLPLANT    R   w FE              iL GD LY RMDY       h    aDIt     P    RA    GL KID   G    F Ek P G   L  M  VDT l
atlarge   ALRN.SLAFE D.ARSKDIED VLILSGDHLY RMDYMDLYRI IGRVGADISI SCIPIDDRRA SDFGLMKIDD KGRVISFSEK PKG.DDLKAM AVDTTILGLS 251
x73367    .......... .......... .......... .......... .......... .......... .......... .......... .......... ..........
x14348    .......... .........G VLILSGDHLY RMDYMDFVQS HRQRDAGISI CCLPIDGSRA SDFGLMKIDD TGRVISFSEK PRG.ADLKEM .......... 70
x62243    .......... .......... .......... .......... .......... .......... .......... .......... .......... ..........
x76136    AVRQFHWLFE D.ARSKDIED VLILSGDHLY RMDYLHFVQS HRQSGADITI SSLPIDDSRA SDFGLMKIDD TGRVMSFSEK PKG.DDLKAM AVDTTVLGLS 250
pcrcode   TV........ .......... .......... .......... .......... .......... .......... .......... .......... .......... 74
x96766    AVRQFTWIFE D.AKNINVEN VLILAGDHLY RMDYMDLLQS HVDRNADITV SCAAVGDNRA SDYGLVKVDD RGNIIQFSEK PKG.ADLKAM QVDTSRLGLS 278
x61187    AVRKFIWVFE D.AKNKNIEN IVVLSGDHLY RMDYMELVQN HIDRNADITL SCAPAEDSRA SDFGLVKIDS RGRVVQFAEK PKG.FDLKAM QVDTTLVGLS 238
x73366    .......... .......... .......... .......... .......... .......... .......... .......... .......... ..........
x78900    AVRQFFWAFE D.SKSKDVEH IVILSGDHLY RMDYMSFWQK HIDTNADITV SCIPMDDSRA SDYGLMKIDH TGRIVHFAEK PKG.SDLTAM QVDTTVLGLS 291
x74982    AVREFIWVFE N.QKNKNVEH IIILSGDHLY RMNYMDFVQK HIDTNADITV SCVPMDDGRA SDFGLMKIDE TGAIIQFAEK PKG.PALKAM QVDTSILGLS 287
x73364    .......... .......... .......... .......... .......... .......... .......... .......... .......... ..........
x14349    .......... .......... .......... .....ELVQK HVDDNADITL SCAPVGESRA SEYGLVKFDS SGRVVQFSEK PKG.DDLEAM KVDTSFLNFA 64
z21969    AVRKFIWVLE DYYKNKSIEH ILILSGDQLY RMDYMELVQK HVDDNADITL SCAPVGESRA SEYGLVKFDS SGRVVQFSEK PKG.DDLEAM KVDTSFLNFA 291
x62242    .......... .......... .......... .......... .......... .......... .......... .......... .......... ..........
x67151    AVRKFIWVLE DYYKHKSIEH ILILSGDQLY RMDYMELVQK HVDDNADITL SCAPVGESRA SEYGLVKFDS SGRVIQFSEK PKG.DDLEAM KVDTSFLNFA 295
x14350    AWRKIIWVLE DYYKNKSIEH ILILSGDQLY RMDYMELVQK HVDDNADITL SCAPVGESRA SEYGLVKFDS SGRVVQFSEQ PKG.DDLEAM KVDTSFLNFA 271
z38111    AVRKFIWVLE DYYKHKAIEH ILILSGDQLY RMDYMELVQK HVDDNADITL SCAPVGESRA SDYGLVKFDS SGRVIQFSEK PKG.AALEEM KVDTSFLNFA 287
s48563    SIRKFIWVLE DYYSHKSIDN IVILSGDQLY RMNYMELVQK HVEDDADITI SCAFVDESRA SKNGLVKIDH TGRVLQFFEK PKG.ADLNSM RVETNFLSYA 284
ALLLARGE      F      d    k  e     s          m                   sc         S                                d
ONLYLRGE      F           k  e     s          m                   sc         S                                d
```

f

```
            301                                                                        351                                                                      400
ALLSMALL    D. RAKe          I  i S  VML LLR   FPgAN      VI G t    G r Q   L d G          e A YN  LGIT KKP PD S     RS  I   P
ONLYSMLL       R             I  i S   M          g        V  G      G r       d            e A       I    K   D      R
  d21272    .......... .......... .......... .......... .......... .......... .......... .......... .......... ..........
  x91736    P.EEAAEKPY IASMG.IYVF KKSVLLQLLN DSYAKANDFG GEIIPSAAK. DHNVVAYPFY GYWEDIGTIK SFFEENLKSC HP..ATFEFY DPQSPIYTSP 379
  x76941    D.DRAKEMPY IASMG.IYVV SKHVMLDLLR DKFPGANDFG SEVIPGATEL GMRVQAYLYD GYWEDIGTIE AFYNANLGIT KKPVPDFSFY DRSSPIYTQP 376
  x96764    D.ERAKEMPY IASMG.IYVV SKHVMLDLLR DKFPGANDFG SEVIPGATEL GLRVQAYLYD GYWEDIGTIE AFYNANLGIT KKPVPDFSFY DRSSPIYTQP 380
  u11281    D.QRAKELPF IASMG.IYVI SKNVMLNLLR EKFPGANDFG SEVIPGATSI GMRVQAYLFD GYWEDIGTIE AFYNANLGIT KKPVPDFSFY DRSAPISTTP 167
  x83498    D.QRAKELPF IASMG.IYVI SKNVMLNLLR EKFPGANDFG SEVIPGATSI GMRVQAYLFD GYWEDIGTIE AFYNANLGIT KKPVPDFSFY DRSAPISTTP 291
  z46756    D.QRAKELPF IASMG.IYVI SKNVMLNLLR EKFPGANDFG SEVIPGATSI GMRVQAYLFD GYWEDIGTIE AFYNANLGIT KKPVPDFSFY DRSAPISTTP 166
  x76940    D.ERAKEMPF IASMG.IYVI SKNVMLDLLR DKFPGANDFG SEVIPGATSI GMRVQAYLYD GYWEDIGTIE AFYNANLGIT KKPVPDFSFY DRSSPIYTQP 372
  x96765    D.ERAKEMPF IASMG.IYVI SKNVMLDLLR DKFPGANDFG SEVIPGATSV GMRVQAYLYD GYWEDIGTIE AFYNANLGIT KKPVPDFSFY DRSSPIYTQP 371
  x83500    D.ERAKEMPY IASMG.IYVI SKDVMLNLLR DKFPGANDFG SEVIPGATSI GLTVQAYLYD GYWEDIGTIE AFYNANLGIT KKPVPDFSFY DRSSPIYTQP 208
  l33648    D.KRAKEMPF IASMG.IYVI SKDVMLNLLR DKFPGANDFG SEVIPGATSL GMRVQAYLYD GYWEDIGTIE AFYNANLGIT KKPVPDFSFY DRSAPIYTQP 385
  x61186    D.KRAKEMPF IASMG.IYVI SKDVMLNLLR DKFPGANDFG SEVIPGATSL GMRVQAYLYD GYWEDIGTIE AFYNANLGIT KKPVPDFSFY DRSAPIYTQP 385
  x55155    D.KRAKEMPF IASMG.IYVI SKDVMLNLLR DKFPGANDFG SEVIPGATSL GMRVQAYLYD GYWEDIGTIE AFYNANLGIT KKPVPDFSFY DRSAPIYTQP 306
  x55650    D.KRAKEMPF IASMG.IYVI SKDVMLNLLR DKFPGANDFG SEVIPGATSL GMRVQAYLYD GYWEDIGTIE AFYNANLGIT KKPVPDFSFY DRSAPIYTQP 306
  l41126    D.KRAKEMPF IASMG.IYVI SKDVMLNLLR DKFPGANDFG SEVIPGATSL GMRVQAYLYD GYWEDIGTIE AFYNANLGIT KKPVPDFSFY DRSAPIYTQP 300
  x78899    D.ERAKEMPF IASMG.IYVI SKDVMLNLLR EQFPGANDFG SEVIPGATSI GLRVQAYLYD GYWEDIGTIE AFYNANLGIT KKPVPDFSFY DRSSPIYTQP 365
  j04960    D.VRAKEMPY IASMG.IYVI SKNVMLQLLR EQFPGANDFG SEVIPGATNI GMRVQAYLYD GYWEDIGTIE AFYNANLGIT KKPVPDFSFY DRSAPIYTQP 343
  m31616    D.VRAKEMPY IASMG.IYVI SKNVMLQLLR EQFPGANDFG SEVIPGATNI GMRVQAYLYD GYWEDIGTIE AFYNANLGIT KKPVPDFSFY DRSAPIYTQP 343
  x62241    ......KYPY IAGMG.IYVI SKHVMLQLLR EQFPGANDFG SEVIPGATST GMRVQAYLYD GYWEDIGTIE AFYNANLGIT KKPIPDFSFY DRSAPIYTQP 93
  x66080    D.ARAKEMPY IASMG.IYVI SKHVMLQLLR EQFPGANDFG SEVIPGATST GMRVQAYLYD GYWEDIGTIE AFYNANLGIT KKPIPDFSFY DRSAPIYTQP 337
  z48562    D.ARAKEMPY IASMG.IYVI SKHVMLQLLR EQFPGANDFG SEVIPGATST GMRVQAYLYD GYWEDIGTIE AFYNANLGIT KKPIPDFSFY DRSAPIYTQP 336
  z48563    D.ARAKEMPY IASMG.IYVI SKHVMLQLLR EQFPGANDFG SEVIPGATST GMRVQAYLYD GYWEDIGTIE AFYNANLGIT KKPIPDFSFY DRSAPIYTQP 377
  x73365    .......YPY IAGMG.IYVV SRDVMLDLLR NQFPGANDFG SEVIPGAPFL GLRVQAYLYD GYWEDIGTIE AFYNANLGIT KKPVPDFSFY DRSAPIYTQP 92
brittle2    D.VRAKEMPY IASMG.IYVF SKDVMLQLLR EQFPEANDFG SEVIPGATSI GKRVQAYL.H GYWEDIGTIA AFYNANLGIT KKPIPDFSFY DRFAPIYTQP 337
  s72425    .......... .......... .......... .......... .......... .......... .......... .......... .......... ..........
 atsmall    D.QRAKEMPF IA........ .......... .......... .......... .......... .......... .......... .......... .......... 353
ALLPLANT            Py IASMG  Yv     l LL         DFG SE  P a      V AY     YWEDiGTi  F  An       P   F FY D   P  T
 atlarge    K.EEAEKKPY IASMG.VYVF KKEILLNLLR WRFPTANDFG SEIIPFSAK. EFYVNAYLFN DYWEDIGTIR SFFEANLALT EHP.GAFSFY DAAKPIYTSR 347
  x73367    .......YPY IAGMG.VYVF KKEILLNLLR WRFPTANDFG SEIILL.AK. EFYVNAYLSN DYWEHIGTIR SFFEANLALT EHP.GAFSFY DAAKPIYTSR 89
  x14348    ..EEAEKKPY IASMG.VYIF KKEILLNLLR WRFPTANDFG SEIIPAAAR. EINVKAYLFN DYWEDIGTIK SFFEANLALA EQP.SKFSFY DASKPMYTSR 165
  x62243    ......KYPY IAGMG.VYIF KKEILLNLLR WRFPTANDFG SEIIPAAAR. EINVKAYLFN DYWEDIGTIK SFFEANLALA EQP.SKFSFY DASKPMYTSR 91
  x76136    P.EEAKEKPY IASIGKVYVF KKDILLNLLR WRFPTANDFG SEIIPASTK. EFCVKAYLFN DYWEDIGTIR SFFRANLALT EHP.PRFSFY DATKPIYTSR 347
 pcrcode    .......... .......... .......... .......... .......... .......... .......... .......... .......... ..........
  x96766    P.QDALKSPY IASMG.VYVF KKDVLLKLLK WRYPTSNDFG SEIIPSAIR. EHNVQAYFFG DYWEDIGTIK SFYDANLALT EES.PKFEFY DPKTPIFTSP 374
  x61187    P.QDAKKSPY IASMG.VYVF KTDVLLKLLK WSYPTSNDFG SEIIPAAID. DYNVQAYIFK DYWEDIGTIK SFYNASLALT QEF.PEFQFY DPKTPFYTSP 334
  x73366    .......YPY IAGMG.VYVF RKEGLLKLLR SSYPTSNDFG SEIIRARRK. LHNVQAFLFN DYWEDIGTIG SFFDANLALT EQP.PKFQFY DQKTPFFTSP 90
  x78900    DLE.AMSNPY IASMG.VYVF RTDVLMELLN RKYPSSNDFG SEIIPSAVG. ESNVQAYLFN DYWEDIGTIK SFFDSNLALT QQP.PKFEFY DPKTPFYTSA 387
  x74982    EQE.ASNFPY IASMG.VYVF KTDVLLNLLK SAYPSCNDFG SEIIPSAVK. DHNVQAYLFN DYWEDIGTVK SFFDANLALT KQP.PKFDFN DPKTPFYTSA 383
  x73364    .......YPY IAGMG.VYCF KTEALLKLLT WRYPSSNDFG SEIIPAAIK. DHNVQGYIYR DYWEDIGTIK SFYEASIALV EEH.PKFEFY DQNTPFYTSP 90
  x14349    .IDDPAKYPY IASMG.VYVF KRDVLLNLLK SRYAELHDFG SEILPRALH. DHNVQAYVFT DYWEDIGTIR SFFDANMSLC EQP.PKFEFY DPKTPFFTSP 160
  z21969    .IDDPAKYPY IASMG.VYVF KRDVLLNLLK SRYAELHDFG SEILPRALH. DHNVQAYVFT DYWEDIGTIR SFFDANMALC EQP.PKFEFY DPKTPFFTSP 387
  x62242    ......KYPY IASMG.VYVF KRDVLLNLLK SRYAELHDFG SEILPRALH. DHNVQAYVFT DYWEDIGTIR SFFDANMALC EQP.PKFEFY DPKTPFFTSP 91
  x67151    .IDDPAKYPY IASMG.VYVF KRDVLLNLLK SRYAELHDFG SEILPRALH. DHNVQAYVFT DYWEDIGTIR SFFDANMALC EQP.PKFEFY DPKTPFFTSP 391
  x14350    .IDDPAKYPY IASMG.VYVF KRDVLLNLLK SRYAELHDFG SEILPRALH. DHNVQAYVFT DYWEDIGTIR SFFDANRALC EQP.PKFEFY DPKTPFFTSP 367
  z38111    TCTLPAEYPY IASMG.VYVF KRDVLLDLLK SRYAELHDFG SEILPKALH. EHNVQAYVFT DYWEDIGTIR SFFDANMALC EQP.PKFEFY DPKTPFFTSP 384
  s48563    .IDDAQKYPY LASMG.IYVF KKDALLDLLK SKYTQLHDFG SEILPRAVL. DHSVQACIFT GYWEDVGTIK SFFDANLALT EQP.SKFDFY DPKTPFFTAP 380
ALLLARGE             Y        v F K   L                   I    .           d         S F    aL       .
ONLYLRGE                      v   K   L                   I    .           d         S F    aL       .
```

FIG. 3. *Continued*

g

```
          401                                               451                                                  500
ALLSMALL      SK LD AD TDSV GE   VIKN KI    VV L  CI SEGAIIED L LM   Y      ad   L  k        I     h   A
ONLYSMLL         LD AD      GE   VIK           L  C   E AIIE  L L           ad      k              h   A
  d21272 .......... .......... .......... .......... .......... .......... .......... .......... .......... ..........
  x91736 ALLPPATVRN CKVTDAIIAQ GSFVSDCTIN NAVIGIRSII GQNCTIQDAL VMGADYYESD DQRATLL..K KGGVPVGIGA NSVITNAIID KNARVGKNVK 477
  x76941 RYLPPSKMLD ADITDSVIGE GCVIKNCKIH HSVVGLRSCI SEGAIIEDTL LMGADYY.ET DADRRFLAAK GG.VPIGIGK NSHIRRAIID KNARIGDDVK 474
  x96764 RYLPPSKMLD ADITDSVIGE GCVIKNCKIH HSVVGLRSCI SEGAIIEDTL LMGADYY.ET DADRRFLAAK GG.VPIGIGK NSHIKRAIID KNARIGDDVK 478
  u11281 RYLPPSKMLD ADVTDSVIGE GCVIKNCKIH HSVVGLRSCI SEGAIIEDSL LMGADYY.ET DADRRLLAAK GS.VPIGIGR NSHIKRAIIH NIARIGNDVK 265
  x83498 RYLPPSKMLD ADVTDSVIGE GCVIKNCKIH HSVVGLRSCI SEGAIIEDSL LMGADYY.ET DADRRLLAAK GS.VPIGIGR NSHIKRAIIH NIARIGNDVK 389
  z46756 RYLPPSKMLD ADVTDSVIGE GCVIKNCKIH HSVVGLRSCI SEGAIIEDSL LMGADYY.ET DADRRLLAAK GS.VPIGIGR NSHIKRAIIH NIARIGNDVK 264
  x76940 RYLPPSKMLD ADITDSVIGE GCVIKNCKIF HSVVGLRSCI SEGAIIEDTL LMGADYY.ET EADKRFLAAK GS.VPIGIGK NSHIKRAIVD KNARIGENVK 470
  x96765 RYLPPSKMLD ADITDSVIGE GCVIKNCKIF HSVVGLRSCI SEGAIIEDTL LMGADYY.ET EADKRFLAAK GS.VPIGIGK NSHIKRAIVD KNARIGENVK 469
  x83500 RYLPPSKMLD ADITDSVIGE GCVIKNCKIH HSVIGLRSCI SEGAIIEDTL LMGADYY.ET DADRKLLAAK GS.VVLGIGQ NSHIKRAIID KNARIGDNVK 306
  l33648 RYLPPSKMLD ADVTDSVIGE GCVIKNCKIH HSVVGLRSCI SEGAIIEDSL LMGADYY.ET DADRKLLAAK GS.VPIGIGK NCHIKRAIID KNARIGDNVK 483
  x61186 RYLPPSKMLD ADVTDSVIGE GCVIKNCKIH HSVVGLRSCI SEGAIIEDSL LMGADYY.ET DADRKLLAAK GS.VPIGIGK NCHIKRAIID KNARIGDNVK 483
  x55155 RYLPPSKMLD ADVTDSVIGE GCVIKNCKIH HSVVGLRSCI SEGAIIEDSL LMGADYY.ET DADRKLLAAK GS.VPIGIGK NCHIKRAIID KNARIGDNVK 404
  x55650 RYLPSSKMLD ADVTDSVIGE GCVIKNCKIH HSVVGLRSCI SEGAIIEDSL LMGADYY.ET DADRKLLAAK GS.VPIGIGK NCHIKRAIID KNARIGDNVK 404
  l41126 RYLPPSKMLD ADVTDSVIGE GCVIKNCKIH HSVVGLRSCI SEGAIIEDSL LMGADYY.ET DAERKLLAAK GSVVPIGIGK NCLYKRAIID KNARIGDNVK 399
  x78899 RYLPPSKMLD ADITDSVIGE GCVIKNCKIH HSVIGLRSCI SEGAIIEDTL LMGADYY.ET DADRKFLAAK GS.VPIGIG. .......... .NARIGDDVK 452
  j04960 RHLPPSKVLD ADVTDSVIGE GCVIKNCKIH HSVVGLRSCI SEGAIIEDSL LMGADYY.ET EADKKLLGEK GG.IPIGIGK NCHIRRAIID KNARIGDNVK 441
  m31616 RHLPPSKVLD ADVTDSVIGE GCVIKNCKIH HSVVGLRSCI SEGAIIEDSL LMGADYY.ET EADKKLLGEK GG.IPIGIGK NCHIRRAIID KNARIGDNVK 441
  x62241 RHLPPSKVLD ADVTDSVIGE GCVIKNCKIH HSVVGLRSCI SEGAIIEDAL LMGADYY.ET EADKKLLAEK GG.IPIGIGK NSHIKRAIID KNARI..... 186
  x66080 RHLPPSKVLD ADVTDSVIGE GCVIKNCKIH HSVVGLRSCI SEGAIIEDTL LMGADYY.ET EADKKLLAEK GG.IPIGIGK NSHIKRAIID KNARIGDNVM 435
  z48562 RHLPPSKVLD ADVTDSVIGE GCVIKNCKIH HSVVGLRSCI SEGAIIEDTL LMGADYY.ET EADKKLLAEK GG.IPIGIGK NSHIKRAIID KNARIGDNVM 436
  z48563 RHLPPSKVLD ADVTDSVIGE GCVIKNCKIH HSVVGLRSCI SEGAIIEDTL LMGADYY.ET EADKKLLAEK GG.IPIGIGK NSHIKRAIID KNARIGDNVM 475
  x73365 RYLPPSKMLD ADVTDSVIGE GCVIKNCKIH HSVVGLRSCI SEGAIIEDSL LMGADYY.ET ATEKSLLSAK GS.VPIGIGK NSHIKRAIID INARIG.... 185
brittle2 RHLPPSKVLD ADVTDSVI.E GCVIKNCKIN HSVVGLRSCI SEGAIIEDSL LMGADYY.ET EADKKLLAEK GG.IPIGIGK NSCIRRAIID KNARIGDNVK 434
  s72425 .......... ..VTDSVIGE GCVIKNCKIH HSVVGLRSCI SEGAIIEDTL LMGADYYAET EADKKLLAEN GG.IPIGIGK NSHIRKAIID KNARIGDNVK 87
 atsmall .......... .......... .......... .......... .......... .......... .......... .......... .......... ..........
ALLPLANT r LPP             i   Gc    C    hs  G rS                GAD Y                G  PiG G  N      iid  Narig  v
 atlarge RNLPPSKIDN SKLIDSIISH GSFLTNCLIE HSIVGIRSRV GSNVQLKDTV MLGADYYKTE AEVAALLAE. .GNVPIGIGE NTKIQECIID KNARVGKNVI 445
  x73367 RNLPPSKIDN SKLIDSIISH GSFLTNCLIE HSIVGIRSRV GSNVQLKDTV MLGADYYQTE AEVAALLAE. .GNVPIGIGE NTKIQEC.ID KNARIG.... 182
  x14348 RNLPPSMISG SKITDSIISH GCFLDKCRVE HSVVGIRSRI GSNVHLKDTV MLGADFYETD MERGDQLAE. .GKVPIGIGE NTSIQNCIID KNARIGKNVT 263
  x62243 RNLPPSMISG SKITDSIISH GCFLDKCRVE HSVVGIRSRI GSNVHLKDTV MLGADFYETD AERGDQLAE. .GKVPIGIGE NTSIQNCIID MN........ 181
  x76136 RNLPPSAIDN SKIVDSIVSH GIFLTNCFVE HSVVGIRSRI GTNVHLKDTV MLGADYYETD AEIRSQLAE. .GKVPLGIGE NTRIKDCIID KNARIGKNVV 445
 pcrcode .......... .......... .......... .......... .......... .......... .......... .......... .......... ..........
  x96766 GFLPPTKIDN SRVVDAIISH GCFLRDCTIQ HSIVGERSRL DYGVELQDTV MMGADYYQTE SEIASLLAE. .GKVPIGIGR NTKIKNCIID KNAKIGKEVV 472
  x61187 RFLPPTKIDN CKIKDAIISH GCFLRDCSVE HSIVGERSRL DCGVELKDTF MMGADYYQTE SEIASLLAE. .GKVPIGIGE NTKIRKCIID KNAKIGKNVS 432
  x73366 RFLPPTKVDK CRILDSIVSH GCFLRECSVQ HSIVGIRSRL ESGVELQDTM MMGADFYQTE AEIASLLAE. .GKVPVGVGQ NTRIKNCIID INARIG.... 184
  x78900 RFLPPTKVDR CKIVDSIVSH GCFLQESSIQ HSIVGVRSRL ESGVEFQDTM MMGADYYQTE SEIASLLAE. .GKVPVGVGQ NTKIKNCIID KNAKIGKDVV 485
  x74982 RFLPPTKVDK SRIVDAIISH GCFLRECNIQ HSIVGVRSRL DYGVEFKDTM MMGADYYQTE CEIASLLAE. .GKVPIGVGP NTKIQNCIID KNAKIGKDVV 481
  x73364 RFLPPTKTEK CRIVNSVISH GCFLGECSIQ RSIIGERSRL DYGVELQDTL MLGADSYQTE SE.SRLLAE. .GNVPIGIGR DTKIRKCIID KNARIG.... 186
  x14349 RYLPPTKSDK CRIKEAIILH GCFLRECKIE HSIIGVPSRL NSGSELKNAM MMGADSYETE DEISRLMSE. .GKVPIGVGE NTKISNCIID MNARIGRDVV 258
  z21969 RYLPPTKSDK CRIKEAIISH GCFLRECKIE HSIIGVRSRL NSGSELKNAM MMGADSYETE DEISRLMSE. .GKVPIGVGE NTKISNCIID MNARIGRDVV 485
  x62242 RYLPPTKSDK CRIKEAIISH GCFLRECKIE HSIIGVRSRL NSGSELKNAM MMGADSYETE DEISRLMSE. .GKVPIGVGE NTKISNSYYD MNARI..... 183
  x67151 RYLPPTKSDK CRIKEAIISH GCFLRECKIE HSIIGVRSRL NSGSELKNAM MMGADSYETE DEISRLMSE. .GKVPIGVGE NTKISNCIID MNARIGRDVV 489
  x14350 RYLPPTKSDK CRIKEAIILH GCFLRECKIE HTAF...SRL NSGSELKNAM MMGADSYETE DEMSRLMSE. .GKVPIGVGE NTKISNCIID MNARIGRDVV 462
  z38111 RYLPPTKSDK CRIKDAIISH GCFLRECAIE HSIVGVPSRL NSGCELKNTM MMGADLYETE DEISRLLAE. .GKVPIGVGE NTKISNCIID MNCQGWKERL 482
  s48563 RCLPPTQLDK CKMKYAFISD GCLLRECNIE HSVIGVCSRV SSGCELKDSV MMGADIYETE EEASKLLLA. .GKVPIGIGR NTKIRNCIID MNARIGKNVV 478
ALLLARGE                    sh   fL               R       l     M       T   E      e. .  V        T I  c
ONLYLRGE                    sh   fL               R       l     M       T   E      e. .           T    c
```

h

```
          501                                                547
ALLSMALL  i  DN VQ    ArET  YF  K    VT I D AL p
ONLYSMLL     N        A ET   F  K     T I D  L p
  d21272  ..........  ..........  ..........  ..........  .......
  x91736  IVNKEG.VQE AAREAEGIYI RSGILVIDKD ALVPTGTTI. ....... 515
  x76941  IINSDN.VQE AARETEGYFI KSGIVTVIKD ALIPSGTVI. ....... 512
  x96764  IINSDN.VQE AARETEGYFI KSGIVTVIKD ALIPSGTVI. ....... 516
  u11281  IINNDN.VQE AARETEGYFI KSGIVTIIKD ALIPSGTII. ....... 303
  x83498  IINNDN.VQE AARETEGYFI KSGIVTIIKD ALIPSGTII. ....... 427
  z46756  IINNDN.VQE AARETEGYFI KSGIVTIIKD ALIPSGTII. ....... 302
  x76940  IINSDN.VQE AARETEGYFI KSGIVTIIKD ALIPSGTVL. ....... 508
  x96765  IINSDN.VQE AARETEGYFI KSGIVTIIKD ALIPSGTVI. ....... 507
  x83500  IINSDN.VQE AARETDGYFI KSGIVTVIKD ALIPSGTVI. ....... 344
  133648  IINKDN.VQE AARETDGYFI KSGIVTVIKD ALIPSGIVI. ....... 521
  x61186  IINKDN.VQE AARETDGYFI KSGIVTVIKD ALIPSGIII. ....... 521
  x55155  IINKDN.VQE AARETDGYFI KSGIVTVIKD ALIPSGIVI. ....... 442
  x55650  IINKDN.VQE AARETDGYFI KSGIVTVIKD ALIPSGIVI. ....... 442
  141126  IINKDN.VQE AARETDGYFI KSGIVTVIKD ALIPSGIVI. ....... 437
  x78899  IINSDN.VQE AARETDGYFI KSGIVTIIKD AMIPSGTVI. ....... 490
  j04960  IINVDN.VQE AARETDGYFI KSGIVTVIKD ALLLAEQLYE VAA.... 483
  m31616  IINVDN.VQE AARETDGYFI KSGIVTVIKD ALLLAEQLYE VAA.... 483
  x62241  ..........  ..........  ..........  ..........  .......
  x66080  IINVDN.VQE AARETDGYFI KSGIVTVIKD ALLPSGTVI. ....... 473
  z48562  IINVDN.VQE AARETDGYFI KSGIVTVIKD ALLPSGTVI. ....... 472
  z48563  IINVDN.VQE AARETDGYFI KSGIVTVIKD ALLPSGTVI. ....... 223
  x73365  ..........  ..........  ..........  ..........  .......
brittle2  ILNADN.VQE AAMETDGYFI KGGIVTVIKD ALLPSGTVI. ....... 472
  s72425  ILNADN.VQE AARETDGYFI KGGIVTVIKD ALLPSGTVI. ....... 125
 atsmall  ..........  ..........  ..........  ..........  .......
ALLPLANT  i n      e a      g  I  sGI     k       gtvi
 atlarge  IANSEG.IQE ADRSSDGFYI RSGITVILKN SVIKDGVVI. ....... 484
  x73367  ..........  ..........  ..........  ..........  .......
  x14348  IANAEG.VQE ADRASEGFHI RSGITVVLKN SVIADGLVI. ....... 302
  x62243  ..........  ..........  ..........  ..........  .......
  x76136  IANSEG.VQE ADRSSEGFYM ASGITVISKN STIPDGTVI. ....... 484
 pcrcode  ..........  ..........  ..........  ..........  .......
  x96766  IANKEG.VQE ADRSEDGFYI RSGITIIMEK ATIEDGTVI. ....... 511
  x61187  IINKDG.VQE ADRPEEGFYI RSGIIIILEK ATIRDGTVI. ....... 471
  x73366  ..........  ..........  ..........  ..........  .......
  x78900  IANTDG.VEE ADRPNEGFYI RSGITIILKN ATIQDGLVI. ....... 524
  x74982  ILNKEG.VEE ADRSAEGFYI RSGITVIMKN ATIKDGTVI. ....... 520
  x73364  ..........  ..........  ..........  ..........  .......
  x14349  ISNKEG.VQE ADRPEEGYYI RSGIVVIQKN ATIKDGTVV. ....... 297
  z21969  ISNKEG.VQE ADRPEEGYYI RSGIVVIQKN ATIKDGTVV. ....... 524
  x62242  ..........  ..........  ..........  ..........  .......
  x67151  ISNKEG.VQE ADRPEEGYYI RSGIVVIQKN ATIKDGTVV. ....... 528
  x14350  ISNKEG.VQE ADRPEEGYYI RSGIVVIQKN ATIKDGTVV. ....... 501
  z38111  HNKQRGRSKS PDRPGRRILI RSGIVVVLKN ATIKDGTVI. ....... 521
  s48563  ITNSKG.IQE ADHPEEGYYI RSGIVVILKN ATINECLVI. ....... 517
ALLLARGE      eG       D           R         n   I d
ONLYLRGE      eG       D           R         n     d
```

FIG. 3. Alignment of the primary structures of ADPGlc PPase proteins from various plants. The sequences were obtained from either GenBank or EMBL nucleotide sequence library, except for *Zea mays brittle* 2 (obtained from L. C. Hannah) and *Arabidopsis thaliana* large and small subunits (B. S. White and J. Preiss, unpublished results, 1998). Alignment of the sequences was done essentially as described by Smith-White and Preiss (1992). The small subunit sequences are shown in the top and the large subunit in the bottom. Also shown are five diagnostic classes that have been formulated for residues in the sequence, which are residues in sequences found only in the small subunit, residues found in all members of the small subunit class sequences, residues found in sequences present only in the large subunit, residues found in all members of the large subunit class, and residues found in all plant ADPGlc PPases.

similarity in sequence between the small subunits isolated from different plants and tissues supports this view. In the case of the large subunit, in which amino acid sequences have less similarity to what is observed for the small subunits, it is possible that the different large subunits lend different regulatory properties for the heterotetrameric ADPGlc PPases of different species and/or tissues. Thus, because the different sequences of the large subunit reflect their occurrences in different plant tissues (e.g., leaf, stem, guard cells, tuber, endosperm, root) (Smith-White and Preiss, 1992), it is possible that these sequence differences render the isolated enzyme from different tissues to have different allosteric properties.

VIII. HYDROPHOBIC CLUSTER ANALYSIS

Preparation of single crystals followed by X-ray diffraction analysis can picture accurately the structure of a protein at a high, atomic resolution. Computing advances have accelerated the process of converting a diffraction pattern into a molecular model. However, crystallization is far from a routine procedure, as the conditions required by a particular protein can only be found by screening a multitude of media (now available commercially) known to favor crystallization. Obtaining a crystal is a hit-or-miss business with no theory; the proteins whose structures have been revealed so far have not been chosen for their interest but because of their propensity to crystallize. Sometimes good crystals can be grown, but they do not diffract because they have very large unit cell dimensions or they decay rapidly in the X-ray beam. Some proteins are easier to crystallize than others, and the ADPGlc PPase from *E. coli* and *Anabaena* are among the more difficult ones. One factor affecting crystallization is the high degree of hydration of the molecule, and so far only small crystals have been obtained, and these were unstable under X-ray diffraction (Mulichak *et al.*, 1988).

Until good crystals have been obtained, there are other avenues for obtaining information about the structure of proteins that are difficult to crystallize. As observed by Kendrew when he solved the structure of the myoglobin, the main driving force for folding water-soluble globular protein molecules is to pack hydrophobic side chains into the interior of the molecule, thus creating a hydrophobic core and a hydrophilic surface. The main chain in the interior is arranged in secondary structures to neutralize its polar atoms through hydrogen bonds. There are two main types of secondary structure: alpha- (α-) helices and beta- (β-) sheets. Protein structures are built up by a combination of secondary structural elements, α-helices, and β-strands. These form the core regions—the interior of the molecule—

and are connected by loop regions at the surface. Schematic diagrams where these structures are highlighted are useful; in addition, α-helices and β-strands that are adjacent in the amino acid sequence are usually adjacent in the tertiary structure.

Hydrophobic cluster analysis (HCA) is a technique that displays the clusters of hydrophobic amino acids present within the primary sequence. It has been used to align amino acid sequences, to predict secondary structures, and to help find similar structures in proteins with a low homology (Lemesle-Varloot *et al.,* 1990). The amino acids are plotted as an α-helix and the representation is duplicated to avoid cutting off clusters with the "wrapping" that occurs when one turn of the helix is completed. In the original method, hydrophobic amino acids are highlighted and encircled to signal the presence of a hydrophobic cluster. This technique was applied to the ADPGlc PPase from *E. coli* (Ballicora *et al.,* 1996) and, since the technique is most useful when homologous proteins are compared, the sequence of the enzyme from a cyanobacteria (*Anabaena*) was also analyzed. The modification by Rost and Sander (1993) of the original technique facilitates the identification of clusters and, in the case of ADPGlc PPases from *E. coli* and *Anabaena,* it stresses the similarities between the two proteins. Proline and glycine are known "breakers" of helices and sheets, and a cluster is not drawn when one of these amino acids is included in it. From the hydrophobic analysis using the profile neural network (PHD) program, it is clear that the ADPGlc PPases from *E. coli* and *Anabaena* are identical in the position of many clusters, and in some others the differences are small. There are some insertions and deletions in the sequence, but they do not alter the general pattern of the clusters because, in these insertions, the analysis shows no buried amino acids. This suggests that the small insertions seen among ADPGlc PPases are not part of the "core" of the protein. Analysis of higher plant ADPGlc PPases show a similar pattern of clusters. For example, even though the homology in amino acid sequence is lower between the enzyme from *E. coli* and the small and large subunits of the potato tuber enzyme, all the clusters present in the bacterial enzyme are also present in both subunits of the plant ADPGlc PPase. This indicates that the ADPGlc PPases from different sources share a common folding pattern, despite a different quaternary structure (heterotetramer in plants, homotetramer in bacteria) and a different specificity for the activator.

If the ADPGlc PPases from different sources have similar three-dimensional structures, the structure of one should help predict the secondary structure of another. The sequence of enzymes from *E. coli* and *Anabaena,* and also from the two subunits of the potato tuber enzyme, were analyzed using the PHD program. One general structure that fits all of these proteins was predicted (Fig. 4). The ADPGlc PPase is an α/β protein, but some parts of it are mainly beta, such as the C-terminal and the domain denoted as 3. To verify whether the model is valid, it was tested against the biochemical data available, including the results of partial proteolysis that is, trypsin treatment of the *Anabaena* and the *E. coli* enzymes (Y. Y. Charng and J. Preiss, unpublished results, 1992) and proteinase K (M. Wu and J. Preiss, unpublished results, 1997) digestion of the *E. coli* enzyme. The peptides obtained by protease treatment were analyzed (Fig. 5). Exposed loops would be more sensitive to proteolysis, and the protease studies, which actually cut in sites predicted as loops by the model, confirm the structure proposed. The only exception is the α-helix predicted near the

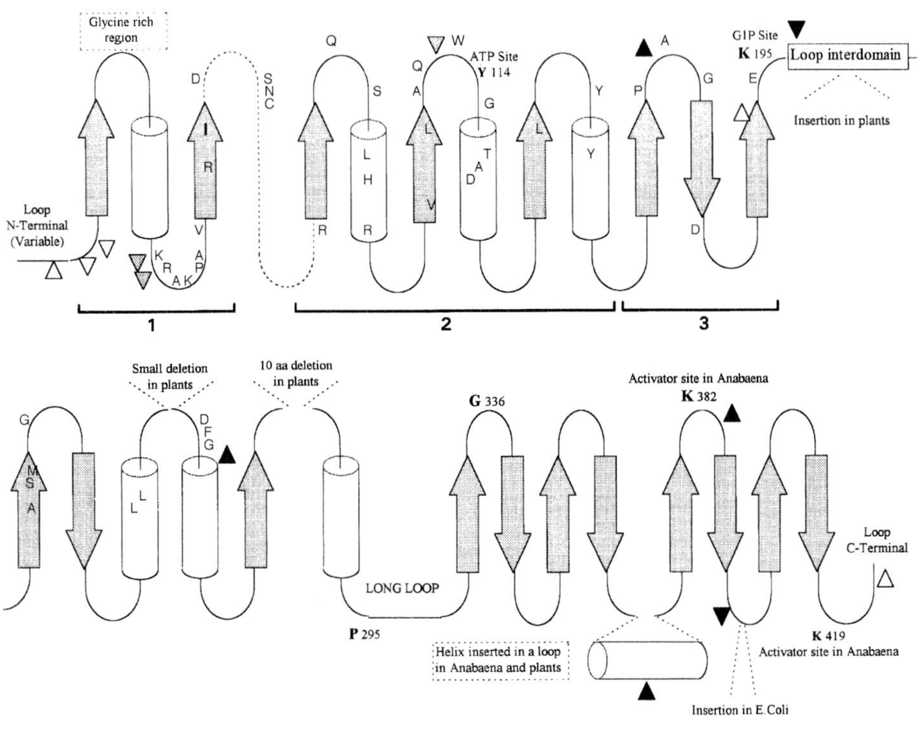

FIG. 4. Profile neural network (PHD) prediction of the secondary structure of the ADPGlc PPase. The structure shown was obtained by applying the program to the sequences of ADPGlc PPases from *E. coli* and *Anabaena,* and the two subunits of the potato tuber ADPGlc PPase. Section 1 contains the Fru-1,6-BP activator site KRAKPAV in a loop as well as R67. Section 2 has the putative ATP binding site, Y114, in a loop area between a β-strand and an α-helix starting at GTAD. The Glc-1-P binding site is also seen in a loop among a series of predicted β-strands. The topology between regions 1 and 2 cannot be ascertained (dotted line).

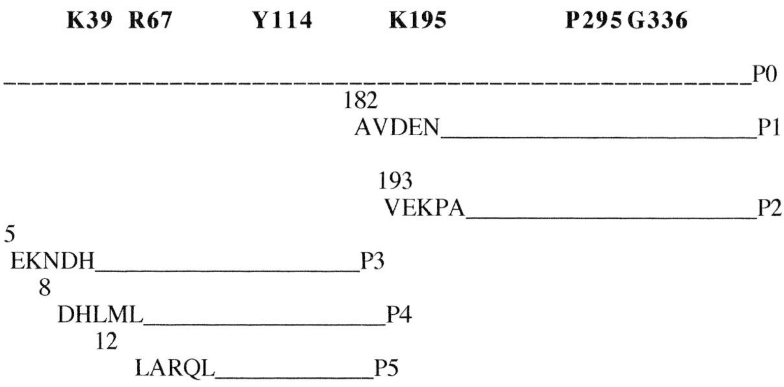

FIG. 5. Controlled digestion of the ADPGlc PPase from *E. coli* by protease K. The full segment P0 represents the sequence of the native enzyme; the amino acid residues known to be important in enzyme function, binding of the substrates, or of the allosteric ligands are shown. Protease K cleaves first the more accessible peptidic linkages 181 to 182 and 192 to 193, inactivating the enzyme and originating the peptides P1 and P2. Cleavage also occurs at the *N*-terminal side of the polypeptide at the 4-5, 8-9, and 11-12 bonds, giving rise to peptides P3, P4, and P5. When the enzyme is incubated with the protease in the presence of ADPGlc, fructose-1,6-bis-P and Mg^{++}, the internal peptide bonds are protected, and only degradation at the *N*-terminal is observed.

C-terminal of the *Anabaena* enzyme. Since this is an insertion (20 aa) that is absent in *E. coli,* and it is not predicted as buried in *Anabaena,* most likely this helix is not part of the core and is part of a loop. It is also worth noting that most of the conserved amino acids known to have roles in the binding of substrates (*E. coli* Y114, K195) and activators (*E. coli* K39, *Anabaena* K382, K419) are located in loops or are very close to loops. The residues P295 and G336 that are involved in areas important for the regulation of the *E. coli* enzyme (Preiss and Romeo, 1989, 1994; Preiss, 1996) are also in loops.

A common supersecondary structure ("motif") seen in nucleotide binding proteins in general (Rossman *et al.,* 1974) is also present in this model—that is, the glycine loop in the domain 1, which would bind the phosphates of the ATP, and the region 2, with three β-sheets and α-helices compatible with a Rossmann fold. It is likely that regions 1, 2, and 3 form a catalytic domain, composed of a typical α/β structure where the substrates bind on the top of the model as depicted in Fig. 4. The prediction of the secondary structure of the ADPGlc PPase in region 1 + 2, is identical to the accepted structure of the oncogenic protein H-Ras (p21), which is used as one of the folding models for nucleotide phosphate binding GTP (Tong *et al.,* 1991).

In region 2, the loops on the N side of the β-sheets (C end of the helices) have no amino acids conserved in all the sequences of the ADPGlc PPases known. This is compatible with the idea that the binding of ATP is located in the other side of the α/β structure. For topological reasons, these loops would not be accessible to the substrate and, as a consequence, evolutionary pressure to conserve the amino acids in these loops is lower than in the loops located at the C end of the β-sheets.

IX. TRANSCRIPTION

Reeves *et al.* (1986) determined the levels of the wheat gliadin and ADPGlc PPase polypeptides, and of their respective mRNAs, using gliadin cDNAs and antibody to the spinach leaf ADPGlc PPase during wheat endosperm development. The mRNA contents for these proteins accumulated coordinately during endosperm development. Gliadin mRNA could be detected at 1% of the maximum level as early as 3 days after flowering (DAF). The mRNA levels for both the gliadins and ADPGlc PPase reached a maximum at about 14 DAF. Thereafter, the mRNA for the ADPGlc PPase decreased whereas the gliadin mRNAs decreased only after 18 DAF. The pyrophosphorylase enzyme increased to a maximum together with its mRNA until 14 to 18 DAF, and then decreased. In contrast, there seemed to be a delay in the expression of the gliadin proteins, and the maximum level was not reached until 31 DAF. Thus, there may be additional levels of control at the translational level since the gliadin proteins were not observed until several days after the appearance of the mRNA. Even though the mRNA levels of the ADPGlc PPase and gliadins appear to be regulated in the same manner, at the translational or posttranslational levels there may be different regulation modes for the two protein families.

The developmental pattern of the ADPGlc PPase gene was determined by Northern and dot blot hybridization analyses (Anderson *et al.*, 1991). The gene is transcribed at the highest level during early development, about 5 to 7 DAF, attaining a level of about 0.2% of the total mRNA, which declines during the later periods of seed development. This pattern of transcription is consistent with the rate of starch accumulation, which is at its highest 7 to 9 DAF (Perez *et al.*, 1975).

The developmental expression of the gene encoding the potato 50-kDa subunit was studied, and the pattern of accumulation of the corresponding mRNA closely followed ADPGlc PPase activity. Thus, the gene appears to be regulated at the transcriptional level for the wheat. It is evident that the regulation of starch synthesis during development in wheat, rice seeds, and potato tuber is similar. There is a close correlation in the activity of ADPGlc PPase and the starch synthetic rate—results that are consistent with the view that gene expression regulates the rate of starch accumulation.

Thus, regulation of ADPGlc PPase, at both the transcriptional level and by allosteric control of the enzyme, modulates the rate of ADPglucose synthesis and starch synthesis.

Northern blot analysis of mRNA isolated from potato leaf, stolon, and tuber against cDNA for the small subunit, indicated that the ADPGlc PPase gene is expressed in the tuber and leaf, but not in the stolon tissue. Thus, regulation at the transcriptional level during tuber development occurs in a tissue-specific manner (Anderson *et al.,* 1990). The size of the mRNA transcripts are 1.8 kb, both in leaf and tuber, on the basis of Northern blot hybridization, suggesting that the same gene may be expressed in both tissues. Conversely, in rice (Krishnan *et al.,* 1986) and wheat (Olive *et al.,* 1989), different genes seem to be encoding the same type of subunit expressed in different parts of the plant (leaf or endosperm). Further analysis is needed to determine if the same or different genes are expressed in different tissues, and whether the situation is different for the large and small subunits and for different species.

In this decade, several authors have reported evidence of changes in ADPGlc PPase and other starch and carbohydrate enzyme expressions brought about by increased availability of sugar (for review, see Koch, 1996). The sugar-inducible enzymes ("feast genes") can be phosphorylase (St. Pierre and Brisson, 1995), ADPGlc PPase (Müller-Röber *et al.,* 1990; Krapp and Stitt, 1995), granule-bound starch synthase, branching enzyme (Kobmann *et al.,* 1991), sucrose synthase (Sus 1) (Müller-Röber *et al.,* 1990; Karrer and Rodriguez, 1992; Koch *et al.,* 1995), invertase (Kobmann *et al.,* 1991), and sucrose-P synthase (Hesse *et al.,* 1995). Examples include those found to be repressible ("famine genes" induced by sugar starvation or depletion), the α-amylase (Karrer and Rodriguez, 1992), and another sucrose synthase isozyme, *sh* 1 (Koch *et al.,* 1992).

X. GENOMIC DNA

Treatment of the rice genomic DNA with *Eco*R1, *Bam*H 1, and *Hin*d III produced two or three bands of DNA fragments ranging from 3 to 5 kb, which hybridized in Southern blots with the rice ADPGlc PPase cDNA. Based on the cDNA copy standards run on the same gel, it was concluded that there are about three gene copies per haploid genome, and the ADPGlc PPase genes are organized in a small family that could be divided into at least two groups on the basis of the restriction fragments obtained (Krishnan *et al.,* 1986).

Using the cDNA clone for the small subunit of the rice ADPGlc PPase (Krishnan *et al.,* 1986) as a probe, the genomic DNA corresponding to the small subunit of ADPGlc PPase was isolated (Anderson *et al.,* 1991) and its structure was determined by nucleotide sequencing. A comparison of the genomic nucleotide sequence with the isolated cDNA sequence revealed a complex gene structure with 10 exons and 9 introns in a size of about 6 kb. The exon sizes are in a range of 99 to 293 base pairs (bp) and the intron sizes range from 84 to 1435 bp. The first intron was the largest, with 1435 bp. The intron splice sites, with the exception of intron 2, contain GT/AG borders and are similar to the splice site consensus sequences (Mount, 1982; Brown, 1986). The intron-2 site did not follow the GT/AG rule (Breathnach and Chambon, 1981) but did show some similarity to the splice site consensus sequences. Thus it is possible, as suggested by Aebi *et al.* (1987), that the overall splice site sequence rather than the particular bases are necessary for correct splicing. The transcription start point is 30 bp downstream of the TATA box and the polyadenylation site was 188 bp downstream of the stop codon. The TATA or Hogness box is a nearly universal sequence, about 25 bp upstream from the transcription start site, reading TATAAAT, and is probably a site of binding for transcription factors.

In Southern blot analysis, only nuclear DNA hybridized with the tuber cDNA corresponding to the potato tuber ADPGlc PPase small subunit, indicating that the gene encoding the enzyme is localized in the nucleus. It is estimated that there are one to two gene copies per haploid genome, and digestion of the potato nuclear DNA with the restriction enzymes *Eco*R1 and *Hin*d III yielded two or three hybridizable fragments totaling 3.6 (*Eco*R1) or 6.7 kb (*Hin*d III) in size. The structure of a genomic clone encoding the analogous rice endosperm small subunit-specific gene has been determined and is almost 6.5 kb in size (Anderson *et al.,* 1991). This rice endosperm gene is interrupted by 9 introns, indicating a structure that is more complex than that of most plant genes. The estimated length of the potato tuber small subunit PPase gene suggests that it may have a complex exon/intron structure, which is a complexity also observed in two other genes involved in starch metabolism—sucrose synthase (Werr *et al.,* 1985) and the granule-bound starch synthase (from maize endosperm; Klösgen *et al.,* 1986), which have 16 and 14 exons, respectively. The multiple introns present in the ADPGlc PPase, and other genes coding for enzymes of starch metabolism, may have a role in gene expression as in the case of the alcohol dehydrogenase gene (Callis *et al.,* 1987).

STARCH SYNTHASES

I. INTRODUCTION

After the synthesis of the glucosyl donor by the ADPglucose pyrophosphorylase (ADPGlc PPase), the next reaction in the starch biosynthetic pathway involves the transfer of the glucosyl moiety of the sugar nucleotide to a maltosaccharide, glycogen, or starch, forming a new α-1,4-glucosidic linkage. In this step, there are some differences between the bacterial and plant systems. In bacteria such as *Escherichia coli,* only one glycogen synthase, encoded by one glycogen synthase gene, has been found (Kumar *et al.,* 1986). Conversely, in every plant tissue studied, more than one starch synthase has been identified (Preiss, 1982a,b, 1991; Preiss and Romeo, 1989; Sivak and Preiss, 1994; Preiss and Sivak, 1996) and they are encoded by more than just one gene. Some starch synthases are bound to the starch and can only be solubilized by α-amylase digestion of the granule, whereas others, designated as soluble starch synthases (SSS), are found in the soluble portion of the extract.

The biochemical and molecular biology characterization of the multiple forms of starch lags behind that of the other enzymes in starch biosynthesis, a problem that can be attributed to the instability of some of the isoforms.

II. SOLUBLE STARCH SYNTHASES

Work with a variety of plant systems has shown that multiple forms of SSS are present. Studies on barley endosperm, pea seeds, wheat endosperm, sorghum seeds, teosinte seeds, spinach leaf, maize endosperm, potato tuber, and rice seed extracts have indicated the presence of at least two major forms of SSS (reviewed in Preiss and Levi, 1980; Preiss, 1988; Preiss and Sivak, 1996), designated as types I and II. In maize leaf (Dang and Boyer, 1988) and castor bean endosperm (Goldner and Beevers, 1989), only one form of starch synthase was found, but since no extensive purification was attempted, the possibility remained that existing multiforms were not separated. Indeed, Downton and Hawker (1973) did find two forms of starch synthase in maize leaf, and thus the issue of the number of forms in maize leaf remains unresolved. It is important to note that Downton and Hawker (1973) found much greater activity in their extracts than that reported by Dang and Boyer (1988), and the possibility of incomplete extraction by the latter remains.

In maize kernels, soluble starch synthase I (SSSI) elutes from anion exchange columns at lower salt concentrations than soluble starch synthase II (SSSII). Although SSSI has been partly purified from maize kernels, SSSII is more unstable and has been more difficult to purify. In our laboratory, the use of several purification steps has resulted in enzymatic fractions with relatively high specific activities (2.6 and 4.2 μmol glucose incorporated per minute per mg protein for SSSI and SSSII, respectively), and these enzymatic fractions were free from amylases and branching enzyme. The apparent affinity for ADPglucose, measured by the K_m, is similar for the two forms (Table I). The maximal velocity of the type I enzyme is greater with rabbit liver glycogen than with amylopectin, and the type II enzyme is less active with glycogen than with amylopectin. Citrate stimulation of the primed reaction is greater for type I than for type II. Both forms can use the oligosaccharides maltose and maltotriose as primers when present at high concentrations. Starch synthase I seemed to have more activity than SSSII with these acceptors (Macdonald and Preiss, 1985).

The lower activity for SSSI with amylopectin as a primer, as compared with glycogen, suggests that SSSI may have a higher preference for the short exterior chains (A-chains) that are more prevalent in glycogen than in amylopectin. The reverse may be true for SSSII, where SSSII may have preference for the longer A-chains and B-chains seen in amylopectin. Dif-

TABLE I

PROPERTIES OF THE SOLUBLE STARCH SYNTHASES FROM MAIZE ENDOSPERM

Property	Starch synthase I	Starch synthase II
Molecular mass (kd)	72	95
Affinity for substrates (K_m)		
ADPGlc	0.1 mM	0.1 mM
Amylopectin	0.16 mg/ml	1.5 mg/ml
Amylopectin (with citrate present)	<1 μg/ml	0.09 mg/ml
Relative activity with different primers (amylopectin = 1)		
Amylopectin + citrate	4.4	1.8
No exogenous primer + citrate	5.8	<0.02
Rabbit liver glycogen	2.1	0.6
1.0 M maltose	1.6	1.2
0.1 M maltotriose	0.9	0.5

ferences were also noted in the apparent affinities with respect to the glucan primer. For example, the K_m for the type I enzyme for amylopectin is nine times lower than that of the type II enzyme. It is worth noting that the type I enzyme is active without added primer in the presence of 0.5 M citrate, whereas the type II enzyme is inactive. Citrate decreases the K_m of amylopectin for both types of enzymes; 160-fold for the type I enzyme and about 16-fold for the type II starch synthase with 0.5 M citrate. SSSI and SSSII enzymes also have different molecular masses based on sucrose density ultracentrifugation: SSSI of maize endosperm is about 70 kDa and SSSII is 95 kDa, and they may be immunologically distinct. Antibody prepared against maize endosperm SSSI showed little reaction with SSSII in neutralization tests. In summary, the two soluble forms of maize starch synthase seem to be distinct on the basis of their physical, kinetic, and immunologic properties, and thus are probably products of two different genes. Because of their different kinetic properties and different specificities with respect to primer activities, they may have different functions in the formation of the starch granule (Macdonald and Preiss, 1985). The properties listed in Table I for the maize isoenzymes are likely to represent the situation in other species and tissues. Both types of synthases can use the oligosaccharides—maltose and maltotriose—as primers at high concentrations. For maltose, the immediate new product was maltotriose; for maltotriose, the product was maltotetraose. Starch synthase I seemed to have more activity with the oligosaccharides than did the type II form. It is worth noting that for most (but not all) plant tissues, starch synthase I elutes from an anion exchange column at lower salt concentrations than starch synthase II.

Native type I and type II enzymes also have different molecular masses based on sucrose density ultracentrifugation (Hawker *et al.,* 1974).

It is possible to study the products formed by the action of the enzymes studied, and this is extremely useful for both starch synthase and branching enzyme. After incubation of the substrates with the enzyme, the product is isolated and debranched by the action of isoamylase (from *Pseudomonas* sp.), which hydrolyses the α-1,6 bonds but not the α-1,4 bonds, yielding relatively short linear chains. The chains obtained are characterized using high-performance anion-exchange chromatography (HPAEC, with a Dionex CarboPac PAI, 250 $\times$ 4 mm column); maltodextrin chains with a DP of up to at least 40 units can be easily separated in less than 50 minutes using as eluting solvent a linear gradient in 150 mM NaOH of 0 to 500 m*M* sodium acetate (Guan *et al.,* 1995). Other useful methods include Smith degradation, measurement of the blue value (BV), determination of carbohydrates by the phenolsulfuric method (Takeda *et al.,* 1983), and beta amylolysis limit.

In rice, three isoforms of SSS were separated by anion exchange chromatography, which, in immunoblot, reacted with antibodies raised to the rice waxy protein (Baba *et al.*, 1993). After affinity chromatography of the active fractions, amino-terminal sequences were obtained for the protein bands of 55 to 57 kD (separated by SDS–PAGE) that cross-reacted weakly with serum raised against the rice waxy protein. It is worth noting that this experimental approach does not exclude the possibility that other soluble starch isoform(s) were present that did not cross-react with the antiserum, and other results from Baba *et al.* (1993) indicate that another SSS isoform of 66 kDa is also present in seed extracts. The same authors isolated cDNA clones coding for the putative SSS from maize from an immature rice seed library in λgt 11, using as probes synthetic oligonucleotides designed on the basis of the amino-terminal amino acid sequences available. The longer insert, of about 2.5 kb, was sequenced and was shown to code for a 1878-nucleotide open reading frame. Comparison with the corresponding amino-terminal sequences led the authors to conclude that the protein is initially synthesized as a precursor carrying a long transit peptide at the amino-acid terminus and that the same gene would be expressed in both seeds and leaves.

III. STARCH SYNTHASES BOUND TO THE STARCH GRANULE

Attempts to elute or solubilize the activity had met with little success until Macdonald and Preiss (1983, 1985) incubated ground maize starch granules with α-amylase and glucoamylase. The solubilized starch synthase activity was chromatographed on an anion exchange column and two peaks of activity were obtained, with 80% of the activity residing with starch synthase I, which eluted from the DEAE-cellulose column at a lower salt concentration than the starch synthase II fraction.

The solubilized, granule-bound enzymes showed apparent affinities for ADPGlc approximately ten times higher than those measured before solubilization (0.96 m*M* for the intact granule activity; F. D. Macdonald and J. Preiss, 1985). The granule-bound enzyme is also active with UDPglucose (Macdonald and Preiss, 1983); in the presence of 0.5 M citrate and amylopectin primer, the granule-bound activity with 1 m*M* UDPGlc exhibits about 7% of the activity seen with 1 m*M* ADPGlc. If the concentration of UDPGlc is raised to 20 m*M*, then the activity is about 73% of that of ADPGlc. Upon digestion of the granule with α-amylase and glucoamylase, the UDPGlc activity essentially disappears. Thus, either the appreciable activity observed with high concentrations of UDPGlc was not solubilized (which would suggest that the UDPGlc activity is catalyzed by a different enzyme than the ADPGlc activity) or was denatured during the amylase treatment, or the ability of the starch-bound starch synthase to use UDPGlc is dependent on the close association of the enzyme with the starch granule. Once solubilized, the granule-associated enzyme is specific for the sugar nucleotide, ADPGlc, and thus is similar to the SSS.

Other properties of the granule-solubilized starch synthases are seen in Table II and can be compared with the SSS. The granule-bound enzymes freed of starch now require a primer for activity. The granule-bound starch synthase II, in contrast to the SSSII, has a higher apparent affinity (lower K_m) for amylopectin than the granule type I enzyme. This is opposite to the soluble enzyme forms. However, as seen for the SSSI, the granule-bound starch synthase I has a higher activity with rabbit liver glycogen than with amylopectin. The starch synthases II, soluble or granule-bound, have less activity with rabbit liver glycogen than with amylopectin. The solubilized granule starch synthase I also has activity without added amylopectin in the presence of 0.5 M citrate, whereas the solubilized granule starch synthase II shows little activity in the presence of citrate and absence of primer. The solubilized granule starch synthases I and II can use oligosaccharide primers as do the soluble starch synthases; Table II shows the activities with respect to maltose and maltotriose. Other maltosaccharides can be used as primers and the products of the reaction observed with maltose, maltotriose, maltotetraose, maltopentaose, maltohexaose, and maltononaose are the maltosaccharides with an additional glucosyl unit (e.g., the product with maltopentaose was maltohexaose).

The molecular weight of the native enzymes can be determined in a number of ways, even in the presence of protein contaminants, by using sucrose density gradients or size exclusion chromatography. The major granule starch synthase, I, has a mass of 61,000, which is different than the

TABLE II

ACTIVITY OF THE STARCH GRANULE-BOUND SYNTHASES WITH ADPGlc AND UDPGlc AS SUBSTRATES: EFFECT OF GRINDING AND SOLUBILIZATION

Enzyme fraction	Starch synthase activity (nmol · min^{-1} · mg^{-1} protein)	
	1 m*M* ADPGlc	20 m*M* UDPGlc
Granules, intact	26	18.9
Granules, ground	114	40
α-amylase–treated ground starch	9.7	<0.02
DEAE-sepharose fractions		
Starch synthase I	47	<0.7
Starch synthase II	74	1.7

molecular mass of the soluble starch synthase I, which is 72,000. Both starch synthases II, soluble or granule-bound, have similar mass; approximately 93,000 (Macdonald and Preiss, 1985).

Antibody prepared against SSSI effectively neutralized SSSI activity, but has little or no effect on either the activities of granule-bound starch synthases I or II or even on SSSII. A result consistent with that obtained with the SSS antiserum is the observation that antibody prepared against the starch granule-bound proteins effectively inhibits granule-bound starch synthase I activity, but has little effect on the SSS (Macdonald and Preiss, 1985).

Classification of starch synthases is primarily based on work done on the maize endosperm enzymes, but work on other species supports, at least qualitatively, similar classifications. Thus, two broad categories of starch synthases can be defined depending on whether the enzymes are soluble or starch granule-bound. Within these categories, two classes of starch synthases I and II can be defined on the basis of their behavior on ion exchange and hydrophobic chromatography. In summary, type I elutes from DEAE-sepharose and aminobutyl-sepharose, at lower salt concentrations, have lower molecular masses, show higher activity with glycogen than with amylopectin, and show higher activities in the absence of primer with citrate or with amylopectin as primer and citrate when compared to the type II class. The two classes are also immunologically distinct. The starch granule-bound enzymes can also be distinguished immunologically from their soluble starch synthase counterparts. There are also differences in their molecular weights and in their kinetic properties. Synthesis of a starch granule has not been obtained in vitro. In vivo, synthesis occurs by deposition on the granule surface by the concerted action of starch synthases and branching enzyme. On centrifugation of a crude extract, starch synthase activity is found associated with the starch granules or in the supernatant. It is assumed that this partition is a result of differences in the structure of the isozymes and/or differences in the role they play in the synthesis of the starch components, amylose and amylopectin. Also, for storage organs such as seeds and tubers, the starch granule is formed and grows for several weeks, and it is likely that different isozymes vary in importance during this period. In maize endosperm there are at least four starch synthases: two soluble (Ozbun *et al.*, 1971) and at least two granule-bound (Macdonald and Preiss, 1985). The number of isoforms may vary with the plant species and the developmental stage, but those that have been studied more carefully seem to have a similar number of isoforms. Indeed, as in the case of the pea embryo, an isozyme of starch synthase—starch synthase II—can exist as a soluble and starch-granule bound isozyme (Edwards *et al.*, 1996). The question remains as to whether the soluble and granule-bound forms are both functional. Indeed, Mu-Forster *et al.* (1996) has reported that in maize endosperm, more than 85% of the starch synthase I protein may be associated with the starch granule. This was determined by using an antibody prepared against the starch synthase, but no evidence was presented to indicate that the protein had starch synthase activity.

The cDNA clones that encode the two isozymes of granule-bound starch synthase of the pea embryo are optimally expressed at different times during development (Dry *et al.,* 1992); although isozyme II is expressed in every organ, isozyme I is not expressed in roots, stipules, or flowers (Dry *et al.,* 1992).

It is worth noting that the understanding of the starch synthases lags behind that of the ADPGlc PPase and the branching enzymes. To cover that ground, it will be necessary to achieve expression of the plant enzymes in *E. coli* so that studies of structure–function relationships can be facilitated.

IV. ISOLATION OF THE WAXY PROTEIN STRUCTURAL GENE

Amylose content determines the degree of translucency of the endosperm (hence the name "waxy"), and it affects the cooking and eating qualities of the grains and the industrial properties of the starch extracted from those grains.

Because of extensive genetic evidence, it is widely accepted that granule-bound starch synthase (GBSS) activity is a function of the protein coded by the *waxy* gene. The final product of the *waxy* locus is a protein of molecular weight 58,000 associated with the starch granule. This protein can be extracted by heating the starch with SDS or by incubating at 37°C with 9 M urea, but these methods are too drastic for the extraction of starch synthase activity. In mutants containing the *wx* alelle, there is virtually no amylose, GBSS activity is very low (Nelson and Rines, 1962; Tsai, 1974; Nelson *et al.,* 1978), and the waxy protein is missing. One of the two GBSS, partly purified from maize kernels (Macdonald and Preiss, 1983, 1985), had a molecular weight as determined by sucrose density gradients of 60,000.

Shure *et al.* (1983) prepared cDNA clones homologous to *Wx* mRNA. In subsequent experiments (Federoff *et al.,* 1983), restriction endonuclease fragments containing part of the *Wx* locus were cloned from strains carrying the ac *wx*-M9, *wx*-M9, and *wx*-M6 alleles to characterize further the controlling insertion elements activator (ac) and dissociation (ds). Excision of the ds element from the certain wx alleles produces two new alleles (S5 and S9) that are encoding the wx proteins having altered starch synthase activities (Wessler *et al.,* 1986). Two of these, S9 and S5, had 53 and 32% of the starch synthase activity, respectively, seen in the normal endosperm. Mutant

S9, with higher starch synthase activity, had 36% of the amylose content observed in the nonmutant endosperm, whereas mutant S5, with an even lower starch synthase activity of 32%, had only 21% of the nonmutant maize amylose content. These data further support the view that the waxy protein is involved in amylose synthesis.

The DNA sequence of the *Waxy* locus of *Zea mays* was determined by analysis of both a genomic and an almost full-length cDNA clone (Klösgen *et al.*, 1986), and the *Waxy* locus from barley has been cloned and its DNA has been sequenced (Rohde *et al.*, 1988). Figure 1 shows the deduced amino-acid sequences from the maize and barley clones and compares them with the amino-acid sequence for the *E. coli* ADPglucose-specific glycogen synthase (Kumar *et al.*, 1986). Of interest is that 13 of the first 27 amino acids in the *E. coli* glycogen synthase are identical to the amino-acid sequences found in the plant enzymes. Moreover, for the three plant starch synthases, 18 of the first 30 amino acids are identical and others may be considered homologous. The sequence starting at residue Lys 15 of the bacterial enzyme, . . . KTGGL . . ., is particularly significant. The lysine in the bacterial glycogen synthase has been implicated in the binding of the substrate, ADPglucose (Furukawa *et al.*, 1990), due to the chemical modification of that site by the substrate analogue, ADP-pyridoxal. The finding of similarity of sequences between the bacterial glycogen synthase and the putative plant starch synthases provides more and stronger suggestive evidence that the waxy gene is indeed the structural gene for the granule-bound starch synthase.

The complete deduced amino-acid sequences of the open reading frames of the *Waxy* genes from maize and barley are known. About 75% identity can be seen in the sequence with respect to amino acids. If functionally similar amino acids are considered, then the homology is about 81%. Thus, these two proteins are similar in sequence and probably carry out the same function in the starch granule. There is, however, very little sequence homology with the bacterial glycogen synthase beyond the *N*-terminal sequence.

Genetically based evidence indicates that the protein product of the *Wx* gene, the waxy protein, is likely to be a GBSS, and biochemical evidence of the identity of the GBSS and the waxy protein has also been obtained (Sivak *et al.*, 1993). Starch-bound proteins from developing maize endosperm were solubilized by digesting the starch with amylases, fractionated by chromatography, and analyzed by electrophoresis and immunoblot (Sivak *et al.*, 1993). In maize endosperm, there are at least four starch synthases: two soluble (Ozbun *et al.*, 1971) and two granule-bound (Macdonald and Preiss, 1985). The number of isoforms may vary with the plant species and the developmental stage, but those that have been studied more carefully

Region I

E.coli glycogen synthase	1MQVLHVCSEMFPLLKTGGLADVIGALP
Potato tuber wx protein	4MNLIFVGTEVGPWSKTGGLGDVLRGLP
Cassava Wx protein	4MNLIFVGAEVGPWSKTGGLGDVLGGLP
Maize Wx protein	5MNVVFVGAEMAPWSKTGGLGDVLGGLP
Barley Wx protein	6MNLVFVGAEMAPWSKTGGLGDVLGGLP
Wheat Wx protein	7MNLVFVGAEMAPWSKTGGLGDVLGGLP
Rice Wx protein	6MNVVFVGAEMAPWSKTGGLGDVLGGLP
Rice Soluble Starch synthase	20RSVVFVTGEASPYAKSGGLGDVCGSLP

	Region II	Region III
E.coli Glycogen synthase	372VPSRFEPCGLTQL	397RTGGLADTV
Potato tuber wx protein	397VPSRFEPCGLIQL	422STGGLVDTV
Cassava Wx protein	398VPSRFEPCGLIQL	423STGGLVDTV
Maize Wx protein	398VTSRFEPCGLIQL	423STGGLVDTV
Barley Wx protein	396VTSRFEPCGLIQL	421STGGLVDTV
Wheat Wx protein	410VTSRFEPCGLIQL	435STGGLVDTV
Rice Wx protein	397VPSRFEPCGLIQL	422STGGLVDTV
Rice Soluble Starch Synthase	372MPSRFEPCGLNQL	397GTGGLRDTV

FIG. 1. Regions of amino acids in the sequence of the *E. coli* glycogen synthase that are conserved in granule-bound starch synthases (also known as *Waxy* proteins) and rice seed-soluble starch synthase. The numbers preceding the sequence indicate the residue number from the putative *N*-terminus in the sequence. The sequence in outline form, KTGGL, has been shown for the *E. coli* glycogen synthase to be involved in binding of the sugar nucleotide substrate.

seem to have a similar number of isoforms. Purification of the starch synthase and branching enzymes in large amounts and to a high specific activity has proved to be difficult, and partly for this reason it has not been possible so far to determine how the enzymes interact to produce the two polysaccharides, amylose and amylopectin, that form the starch granule. Starch extracted from developing pea embryos contained starch synthase activity that was associated with the waxy protein, and the molecular weight of the pea starch synthase is about 59,000, as determined by ultracentrifugation in sucrose density gradients. A pea GBSS preparation displayed a relatively high specific activity (more than 10 μmol glucose incorporated per minute per mg protein). When this enzymatic fraction was subjected to SDS polyacrylamide gel electrophoresis and was followed either by protein staining or immunoblot, only the *Wx* protein was visible (Sivak *et al.,* 1993).

Thus, the biochemical examination of starch synthase present in starch granules from two species, maize and pea, strengthens the genetic evidence supporting the role of the *Wx* protein as a GBSS with a major role in the determination of the amylose content of starch. It is still not clear, however, how the loss of starch synthase present in the starch granule causes the disappearane of amylose. It is possible that the interior of the granule is devoid of branching enzyme or, if branching enzyme is in the granule itself, it is not appreciably active. The presence of an active, chain-elongating enzyme (i.e., starch synthase), without an active branching enzyme present (in the presence or absence of some debranching activity), could lead to amylose formation. For a discussion of how the primer is formed, see the chapter, "Open Questions and Hypotheses in Starch."

The starch synthase isozymes in maize endosperm have different molecular masses. The GBSS isozyme I has a molecular mass of 60,000; GBSSII is 95,000. The SSSI has a molecular mass of 72,000, and SSSII is 95,000. Mu *et al.* (1994) have reported the molecular mass of maize endosperm SSSI as 76,000, which is similar to the value reported previously for SSSI (reviewed in Sivak and Preiss, 1994; Preiss and Sivak, 1996). These molecular mass values for the starch synthases are all higher than that of the *E. coli* glycogen synthase with a molecular weight of 52,000 (Kumar *et al.,* 1986).

There is a report that in pea embryo, some of the SSSs may also be bound to the starch granule (Denyer *et al.,* 1993) and that in maize endosperm, some of SSSI adheres to the starch granule (Mu *et al.,* 1994). The conclusions in the pea embryo study (Denyer *et al.,* 1993) are based on positive immunoblots obtained after electrophoresis of the SSS with antibody prepared against the GBSS, and also on the similarity of the amino-acid sequence of three peptides obtained from protease SV8 digests of the SSS. This clearly shows there is a close relationship between SSSII and GBSSII, but does not indicate that they are identical proteins. It is also not clear how much of the SSSII activity is present as granule-bound activity and how much is soluble activity. It is also not surprising that some of the SSSII is present as starch–granule-bound, as SSSII does have an affinity for its substrate, starch.

The evidence for the maize study rests on the observation of a positive immunoblot with antibody prepared against a 76 kDa protein obtained from the starch granule with the SSSI on electrophoresis; the antibody also neutralized the SSSI activity. Since SSSI has affinity for the granule, one would expect to have some of the SSSI protein bound to the granule, and the question is how much is bound and whether the binding is similar to the binding of the GBSS to the starch granule. There is no question that in maize, the GBSSI is immunologically distinct from the SSSI (reviewed in Sivak and Preiss, 1995; Preiss and Sivak, 1996).

The amino-acid sequences of the bacterial glycogen synthase and the plant starch synthases have been compared (Fig. 1). There are three regions of high conservation and at least one of them is involved in binding the substrate, ADPglucose (Furukawa *et al.,* 1990, 1993). This region (region I) is at the *N*-terminal. The possible functions for regions II and III are not known. However, the high degree of conservation of amino acids in region II (only one or two amino acids, at most, differ from those in the sequence of glycogen synthase of *E. coli*) suggesting that this is likely to be an important site. In region III, all the GBSSs are identical with respect to the amino-acid sequence, whereas the *E. coli* sequence differs in only two of the nine amino acids: Arg for Ser and Ala for Val. The SSS has Gly for that Ser and an Arg residue instead of Val. In addition, Lys residue 277 of the *E. coli* glycogen synthase is also involved in catalysis (Furukawa *et al.,* 1994) and is also conserved in the GBSS and SSS. Many questions remain with respect to protein–structure–function relationships among the three types of α-1,4 glucan synthases and the primer binding site and amino-acid residues involved in catalysis. In rice seed, there is no question that the SSSs are different than the granule-bound starch synthases in that there are only 29 to 37% identities with the rice GBSSs (Baba *et al.,* 1993).

Thus far, the only α-1,4 glucan synthase reported to be overexpressed with high activity is the *E. coli* enzyme. This system should be further exploited with respect to the methodologies of chemical modification, site-directed mutagenesis, and attempts to determine its three-dimensional structure.

V. STUDIES OF *Chlamydomonas reinhardtii* MUTANTS

It is to be expected that different enzymes must have different functions in the synthesis of the starch components, amylose and amylopectin, and this is what the work with *Chlamydomonas reinhardtii* of Ball and his collaborators (1991) indicates (Delrue *et al.,* 1992; Fontaine *et al.,* 1993; Maddelein *et al.,* 1994). These authors showed that SSSII may be involved in the synthesis of the intermediate sized chains of amylopectin (Fontaine *et al.,* 1993) and that GBSS is not only involved in amylose synthesis, but also in amylopectin synthesis (Maddelein *et al.,* 1994). The process to follow is to separate the various granule-bound and soluble starch synthases from each other and to examine their properties. The properties that must be studied are their specificities (their chain-lengthening properties), to what chains they prefer to transfer glucosyl residues (the A-, B1-, B2-, B3-, or B4-chains of amylopectin; Hizukuri, 1986), as well as the optimal length of glucosyl residues they can synthesize efficiently. Amylose synthesis depends on the concentration of ADPGlc, as GBSS has a high K_m for the substrate as compared to the soluble starch synthases (Van den Koornhuyse *et al.,* 1996). The phosphoglucomutase (PGM)-deficient mutants can make amylopectin but not amylose, as shown by detailed structure studies of the starch accumulated by the algae (Libessart *et al.,* 1995), even though the algae have GBSS. A similar structure effect can be seen when the algae have defective ADPGlc PPase (Van den Koornhuyse *et al.,* 1996).

Ball and associates (1991) have isolated various mutants of *Chlamydomonas* that are deficient in starch synthase activities. These are a GBSS-deficient mutant (Delrue *et al.,* 1992), an SSSII-deficient mutant (Fontaine *et al.,* 1993), and a double mutant—that is, a mutant deficient both in GBSS and in SSSII (Maddelein *et al.,* 1994). The SSSII mutant had only 20 to 40% of the wild-type starch content, and the amylose fraction of the starch increased from 25 to 55%. This mutant also had a modified amylopectin with an increased amount of short chains of DP 2 to 7, and a decrease of intermediate size chains of DP 8 to 60. This suggests that the SSSII is involved in the synthesis or maintenance of the intermediate size chains (mainly B-chains) in amylopectin. The higher amylose content could be explained because of the failure of the SSSII mutant to make extended chains.

The double mutants, defective in SSSII and GBSS (Maddelein *et al.,* 1994), had a starch content of only 2 to 16% of the wild-type. The severity of the GBSS defect of the double mutant dictated the amount of starch present in the double mutant, with an almost null mutant having little starch. The authors suggest that GBSS is important for synthesis of the internal structure of the amylopectin, and the effect of GBSS deficiency is worsened by the diminished SSSII activity.

These studies, using *Chlamydomonas* mutants, provide evidence for the involvement of the GBSS, not only in amylose, but also in amylopectin synthesis, and suggest that a function for SSSII is in the synthesis of the intermediate size B-branch chains in amylopectin.

FURTHER READINGS

These sources provide additional in-depth coverage of this topic. For complete reference, please see the Reference section at the end of the book.

Neuffer, M. G., Coe, E. H., and Wessler, S. R. (1997)

BRANCHING ENZYMES

I. INTRODUCTION

The third reaction in the starch biosynthetic pathway is catalyzed by the 1,4-α-D-glucan: 1-4-α-D-glucan 6-glycosyl-transferase (branching enzyme, BE; E.C. 2.4.1.18), responsible for the synthesis of the α-1,6 linkages found in amylopectin and in glycogen. In *E. coli* or other bacteria, only one branching enzyme and one gene are present, whereas in plants, genetic and biochemical evidence indicates there can be two or more forms of branching enzyme. The challenge is to identify the role of each multiform in the synthesis of the complex structure of starch and to understand how differences in enzyme structure determine the specificities in acceptor and branching modes. As usual with plant enzymes, this task is made more difficult by the presence of proteases in crude extracts, and also by the fact that proteolyzed enzymes can still have activity even when their specific properties have been altered in an irreversible manner.

II. ASSAY

Several assays (Fig. 1) are available to measure branching enzyme activity. The iodine assay is based on the decrease in absorbance of the glucan–iodine complex (Krisman, 1962) resulting from the branching of amylose or amylopectin by the enzyme, provided that α-amylases are absent from the enzyme preparation or their activity is greatly reduced by adjusting the assay conditions.

The phosphorylase-stimulation assay (Hawker *et al.*, 1974) is based on the stimulation of the "unprimed" (without added glucan) phosphorylase activity (Fig. 2) of the phosphorylase from a rabbit muscle, as the branching enzyme present in the assay mixture increases the number of nonreducing ends available to the phosphorylase for elongation. It was made more sensitive by using ^{14}C-labeled glucose-1-P and measuring its incorporation into starch, rather than by measuring the formation of P_i.

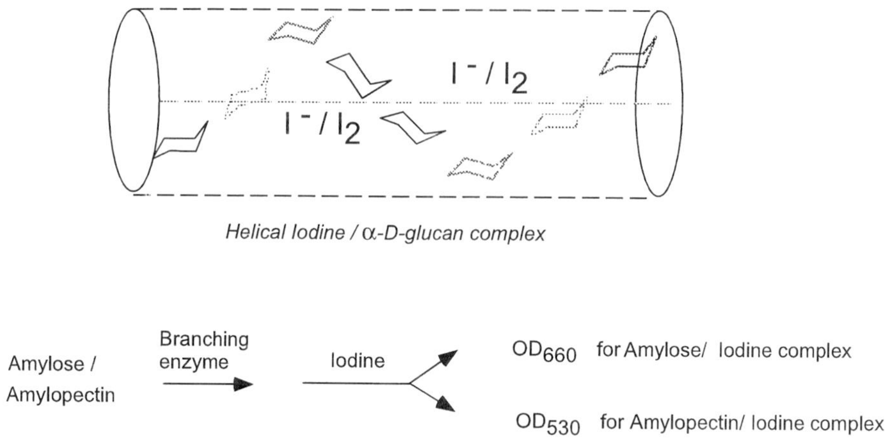

FIG. 1. Iodine staining assay. The substrate, amylose or amylopectin, is branched by the branching enzyme. The assay is terminated by adding iodine/iodide reagent. Amylose and amylopectin form complexes with the iodine/iodide reagent, whose peak of absorbance varies depending on the polymer, and the wavelengths used are 660 nm for amylose/iodine complex and 530 nm for the amylopectin/iodine complex. Branching of the substrate decreases the absorbance of the complex, a fact used to quantify the activity of branching enzyme. Figure reprinted with permission from Binderup (1997).

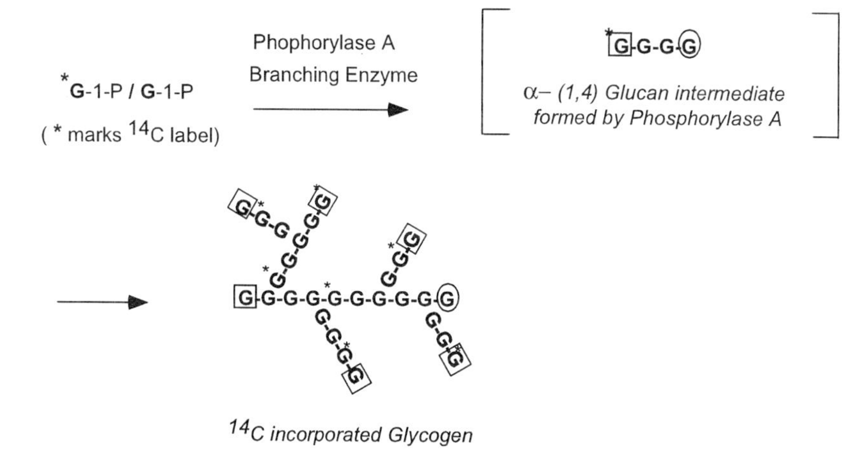

FIG. 2. Phosphorylase stimulation assay of branching-enzyme activity. The phosphorylase a elongates an endogenous primer from the nonreducing end [G]. The branching enzymes originate new branch points, which constitute additional nonreducing terminals, which phosphorylase A can elongate further. The elongated polymer contains ^{14}C-labelled glucose (*G). The only reducing end in the molecule is indicated by the Ⓖ. Figure reprinted with permission from Binderup (1997).

The new branching-linkage assay ("BL assay"; Takeda *et al.,* 1993) is the only one that measures the number of events catalyzed by the branching enzyme (rather than an indirect effect of its action as in the two assays described in the preceding) (Fig. 3). The enzyme fraction is incubated with the substrate, $NaBH_4$-reduced amylose, the reaction is then stopped by boiling, and the product is then incubated with *Pseudomonas* isoamylase for debranching. Finally, the reducing power of the oligosaccharide chains transferred by the enzyme is measured. Reduced amylose is used as the substrate to eliminate or minimize the reducing power of the amylose itself.

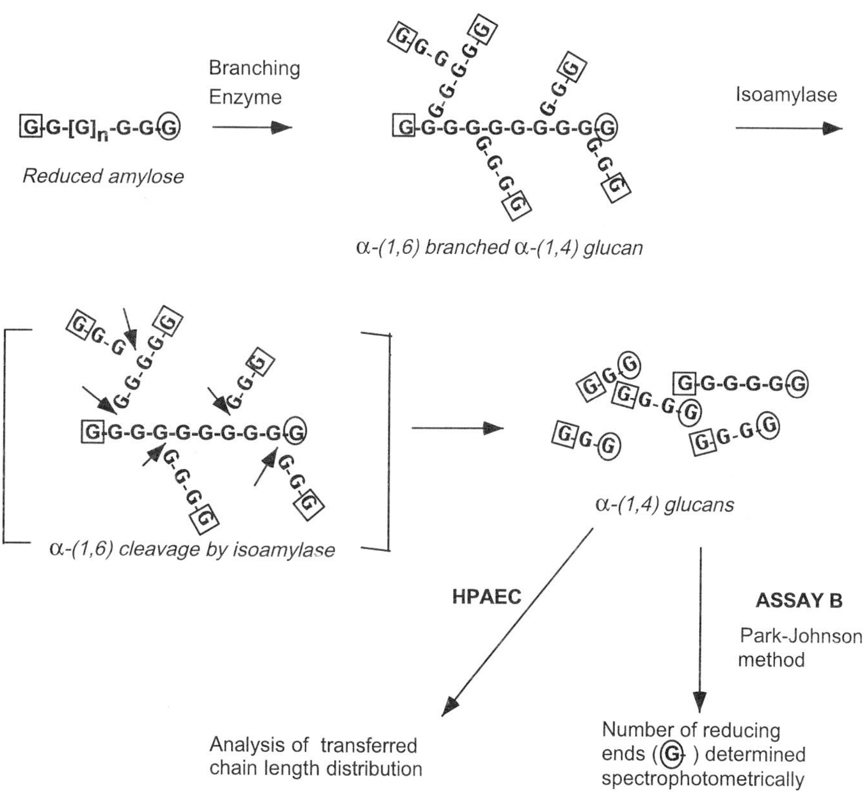

FIG. 3. The mostly linear substrate is branched by branching-enzyme during the assay. After termination of the assay, the now branched polymer is debranched using pure isoamylase, resulting in the liberation of one reducing terminal per chain. In the last step, the reducing terminals are quantified by the Park–Johnson method. Alternatively, the distribution of transferred chains can be determined using HPAEC. Nonreducing terminals are denoted by ▢G, reducing ends are denoted by ⓖ. Figure reprinted with permission from Binderup (1997).

The branching-linkage assay is the most quantitative assay for the branching enzyme, but amylolytic activity interferes most with this assay; the phosphorylase-stimulation assay is the most sensitive and the iodine assay is not very sensitive but allows the testing of branching enzyme specificity with various α-1,4-dextrins, providing information on the possible role of the different branching enzyme isoforms. For complete characterization, it may be best to employ all three assays when studying the properties of the branching enzymes. Above all, however, if reliable information is being sought, the branching enzymes must be purified to the extent that all starch degradative enzymes are eliminated before studying the properties of branching enzymes.

III. PURIFICATION OF BRANCHING ENZYME MULTIFORMS

Multiple forms of branching enzyme have been found in many plants, for example, spinach leaf, maize endosperm, maize leaf (Dang and Boyer, 1988), castor bean (Goldner and Beevers, 1989), developing seeds of pea and sorghum, and teosinte seeds (for reviews see Preiss, 1988). Smyth (1988) resolved two or three peaks of Indica rice starch-branching enzyme on an anion-exchange column. The properties of these fractions are similar to those reported for other plants, except that their molecular weights were determined to be only 40,000.

In maize endosperm, there are three BE isoforms (Singh and Preiss, 1985; Guan and Preiss, 1993), and reports on other tissues are consistent with the presence of more than one isoform. A procedure described by Boyer and Preiss (1978a,b) and by Singh and Preiss (1985), which resulted in enzyme fractions with high specific activity but with traces of amylases, was further improved in our laboratory (Guan and Preiss, 1993). The purification steps involve homogenization, ammonium sulfate precipitation, and chromatography on DEAE-sepharose ("fast-flow"), in which the three isoforms are first separated. BEI is further purified by chromatography on ω-aminodecyl agarose, and by FPLC using MonoQ and Superose 12. Further purification of both BEIIa and BEIIb requires chromatography on ω-aminooctyl agarose and MonoQ. The resulting preparation of BEIIa requires chromatography on BioGel P10 to remove contaminating carbohydrate. Monoclonal antibodies were prepared against BEI and a mixture of BEs IIa and IIb. Singh and Preiss (1985) concluded that although some homology exists among the three starch BEs, there are major differences in the structure of BEI when compared with BEIIa and BEIIb, as shown by its different reactivity with some monoclonal antibodies, and there are differences in amino-acid composition and in proteolytic digest maps. It was also concluded (Singh and Preiss, 1985) that BEs IIa and IIb are similar and possibly are the products of the same gene. Further evidence (Fisher *et al.,* 1993, 1996a), however, suggests that BEIIa and BEIIb may, after all, be products of two different genes (see the later section, "How Many Genes for Three Maize-Branching Enzymes?").

Because of the difficulties involved in separating the isoenzymes among themselves and from other enzymes such as amylases, it is not yet clear what the best acceptors are and what the products are for each isoenzyme, but some characterization of the enzymes has been done (e.g., K_m for α-glucans such as amylopectin, different animal glycogens, and maltosaccharides). Characterization of the products has been minimal, but some progress has been made in our laboratory (see the next section, "Mode of Action").

In potato tubers, Vos-Scheperkeuter *et al.* (1989) purified a single form of branching activity of molecular mass 79,000, confirming the previous work of Borovsky *et al.* (1975). Antibodies were prepared to the native potato enzyme and it was found they reacted strongly only with maize enzyme I and weakly with IIb. In neutralization tests, they inhibited the activities of both the potato tuber-branching enzyme and of maize-branching enzyme I. Antibodies prepared against denatured potato-branching enzyme reacted with all forms of denatured maize- and potato-branching enzyme in immunoblots, but not with the native enzyme forms. It was concluded that the potato-branching enzyme shows a high degree of similarity to the maize-branching enzyme form I and, to a lesser extent, to the other forms of maize-branching enzyme.

IV. MODE OF ACTION

Some subtle differences among BEI and BEII of maize, in the K_m for several primers, and their response to citrate have been described (see Preiss, 1991). More recently, Takeda *et al.* (1993) have analyzed the branched products made from amylose by each BE isoform. This was done by debranching the products of each isoform using isoamylase, followed by gel filtration.

BEIIa and BEIIb are similar in their affinity for amylose and the properties of the products (Tables I and II). When presented with amyloses of different average chain lengths, all the BEs have higher activity with the longer chain amylose. BEI can still catalyze the branching of an amylose of average chain length (c.l.) of 197 with 89% of the activity shown for the c.l. 405. The activity of BEII drops sharply with a reduction in c.l. The action of BEIIa and BEIIb results in the transfer of chains shorter than

TABLE I

SPECIFIC ACTIVITIES (UNITS/MG OF PROTEIN) OF BRANCHING ENZYME ISOFORMS AS MEASURED USING DIFFERENT ASSAY METHODS[a,b]

BE	BEI	BEIIa	BEIIb
(a) Phosphorylase stimulation	1332	795	927
(b) Branching-linkage assay	2.4	0.32	0.33
(c) Iodine stain assay			
(c_1) Primer: amylose	574	29.5	53
(c_2) Primer: amylopectin	47	59	105
Ratio of activity			
a/b	555	2484	2809
a/c_1	2.3	27	18
a/c_2	49.8	13.5	8.8
c_2/c_1	0.03	2	2

[a] Data from Guan and Preiss (1993).

[b] For this experiment, it is not essential to use enzyme purified to homogeneity, hence the relatively low specific activities. The enzymatic fraction, however, must be free of amylases or contaminating branching activities.

BEI. The action of BEI, BEIIa, and BEIIb on amylopectin has also been studied by Guan and Preiss (1993). Of the three isoforms, BEI had the highest activity in branching amylose and its rate of branching amylopectin

TABLE II

PROPERTIES OF THE PRODUCTS OF THE ACTIVITY OF BRANCHING ENZYME ISOFORMS ON AMYLOSE OR AMYLOPECTIN[a]

α-Dextrin + branching enzyme	λ_{max} (nm)	Decrease in absorbance (%)	β-amylolysis limit (%) Before	β-amylolysis limit (%) After isoamylolysis
Amylose, no BE	639	77	100	—
Amylose + BEI	530	89	44	100
Amylose + BEIIa	593	46	48	99
Amylose + BEIIb	600	47	45	98
Amylopectin, no BE	530	—	60	100
Amylopectin + BEI	495	54	53	100
Amylopectin + BEIIa	491	57	46	98
Amylopectin + BEIIb	489	58	42	99

[a] The substrates, either 1 mg/ml amylose (c.l. 405) or amylopectin (c.l. 21), were incubated with 5 units/ml maize-branching enzyme (measured as by stimulation of phosphorylase *a;* Guan and Preiss, 1993). The decrease in absorbance by the iodine/glucan complex was measured at 660 nm.

was less than 5% of that for amylose. In contrast, the BEIIa and BEIIb isoforms branched amylopectin at twice the rate they branched amylose, and catalyzed branching of amylopectin at six times the rate observed for BEI. The assay used was iodine staining (Boyer and Preiss, 1978a,b). These results are consistent with those of Takeda *et al.* (1993) in suggesting that BEI catalyzes the transfer of longer branched chains and that BEIIa and IIb catalyze the transfer of shorter chains. Thus, it is possible that BEI may produce lightly branched polysaccharides, which then serve as substrates for enzyme complexes of branching enzyme II isoforms and starch synthases to synthesize amylopectin. Branching enzyme II isoforms may play a major role in forming the short chains present in amylopectin (Guan and Preiss, 1993).

The study of the reaction products showed that the action of BEIIa and BEIIb resulted in the transfer of chains shorter than those transferred by BEI. The action of the isoforms on amylopectin has been studied by Guan and Preiss (1993) and, of the three isoforms. BEI had the highest activity (using the iodine assay) on amylose, and its rate of branching amylopectin was less than 5% of that with amylose. In contrast, the BEIIa and BEIIb isoforms branched amylopectin at twice the rate of amylose, and catalyzed the branching of amylopectin at six times the rate observed for BEI. In short, the experimental evidence suggests that BEI may be more involved in producing the more interior (B-) chains of the amylopectin, whereas BEIIa and BEIIb would be involved in forming the exterior (A-) chains.

V. HOW MANY GENES FOR THREE MAIZE-BRANCHING ENZYMES?

In neutralization tests, antibody raised against form I inhibits BEI but not BEIIa or BEIIb, and antibody prepared against IIa and IIb inhibits IIa and IIb but not BEI, confirming earlier reports by Fisher and Boyer (1983). However, when an enzyme-linked immunoadsorbent assay (ELISA) was used, some reaction of BEIIa and BEIIb was detected when using antiserum raised against BEI (Singh and Preiss, 1985). The antisera prepared with IIa and IIb, however, reacted only to a very small extent with BEI. This was also observed in using purified monoclonal antibodies raised against BEIIa and BEIIb (Singh and Preiss, 1985). Three monoclonal antibodies could react with all three branching enzymes in the ELISA assay, whereas three other lines produced monoclonal antibodies specific for forms IIa or IIb (Fig. 4). Neutralization of enzyme activity of all three branching enzymes was seen with the antibody that reacted with all three in the ELISA test, whereas the monoclonal antibody that reacted with only

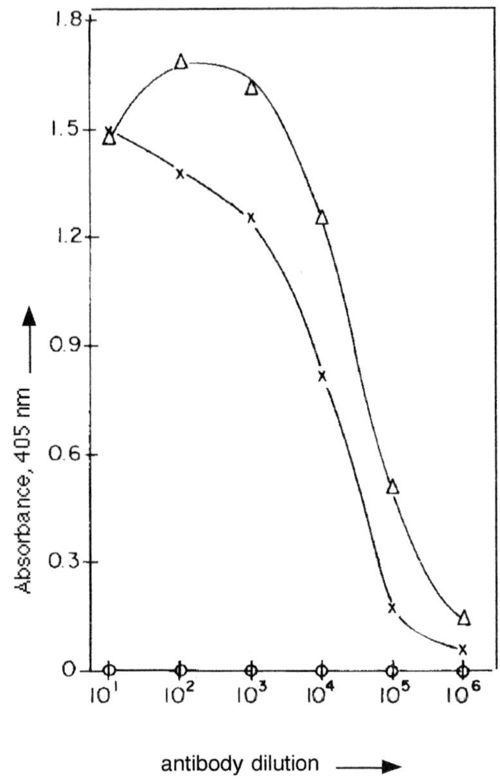

FIG. 4. Monoclonal antibodies obtained from different cell lines (only one cell line is illustrated as an example) reacted with both BEIIa and BEIIb, but did not react with BEI. ○, BEI; x, BEIIa; △, BEIIb. Figure reprinted with permission from Singh and Preiss (1985).

IIa and IIb inhibited their activities in neutralization tests but did not inhibit enzyme I activity.

Amino-acid composition studies also indicate major differences between BEI and the other two forms, whereas only minor differences are observed between IIa and IIb (Singh and Preiss, 1985). High-performance liquid chromatography (HPLC) patterns of a BEI peptide digest (whether trypsin or chymotrypsin were used) are very different in both the number of peptides and their retention times than those obtained with BEIIa or BEIIb digests. The digest patterns produced by trypsin are shown in Fig. 5. In contrast, no differences in HPLC are observed between the peptide maps of BEIIa and BEIIb forms using either trypsin (Fig. 5) or chymotrypsin (Singh and Preiss, 1985). It was concluded that some homology exists among the three starch-branching enzymes, but major differences exist between

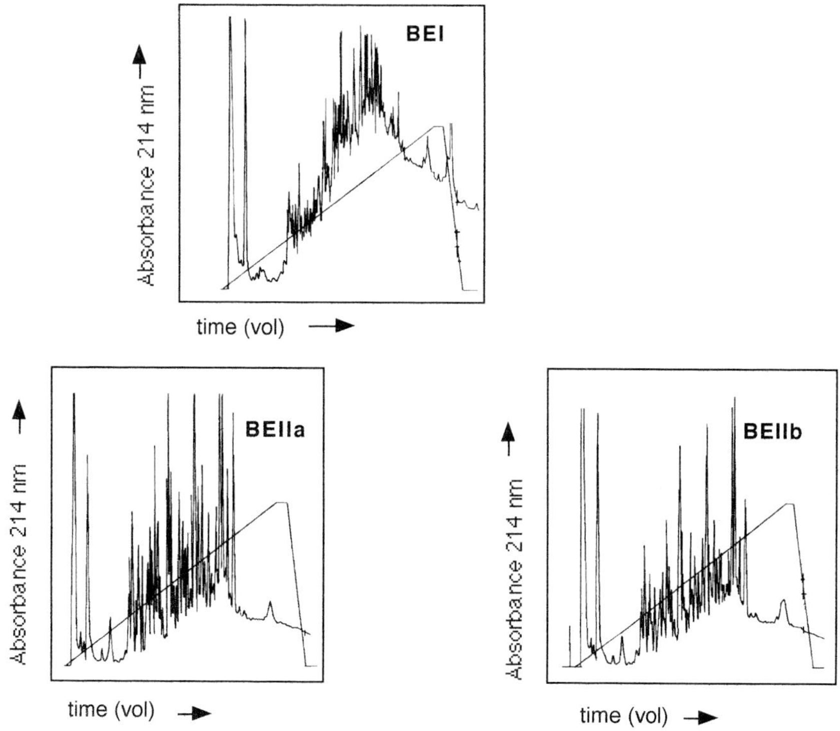

FIG. 5. Tryptic peptide maps of the maize-branching enzymes showing the similarity between the pair BEIIa/BEIIb and differences between the pair and BEI. Approximately 1.25 nmol of peptide digest was chromatographed on a reverse-phase column. Figure reprinted with permission from Singh and Preiss (1985).

the structure of branching enzyme I when compared to the pair IIa and IIb, as shown by the absence of reaction with some monoclonal antibodies, differences in amino acid composition, and differences in their proteolytic digest maps. There were significant differences between BEI and the pair BEIIa and BEIIb, but these two are similar in their reactions to polyclonal and monoclonal antibodies, molecular weight, K_m for substrates, and so on. In maize endosperm (Fisher *et al.,* 1993; Guan *et al.,* 1994a,b) and in rice seed (Mizuno *et al.,* 1992, 1993; Nakamura and Yamanouchi, 1992), two different cDNAs were isolated encoding two branching enzyme isozymes, BEI and BEII.

Results obtained with the maize endosperm mutant, amylose extender (*ae*), suggested that *Ae* was the structural gene for either BEII or BEIIb (Boyer and Preiss, 1978b, 1981; Preiss and Boyer, 1980). BEI levels were not affected. In gene dosage experiments, Hedman and Boyer (1982) showed that there is a nearly linear relationship between increasing the dosage of the dominant *Ae* allele and BEIIb activity. Since the separation of form IIa from IIb was not optimal, it was still possible that the *Ae* locus was also affecting the level of IIa.

In short, the evidence accumulated by 1982 suggested that only two genes were coding for branching enzymes, and that the slight differences between BEIIa and BEIIb could be attributed to posttranscriptional processing. However, on the basis of the analysis of 16 isogenic lines that have independent alleles of the maize *amylose extender* (*ae*) locus, Fisher *et al.* (1996a) suggested that BEIIa and BEIIb are encoded by separate genes; the BEIIb enzyme would be encoded by the *Ae* gene. They isolated a cDNA clone labeled *Sbe* 2b with a predicted amino-acid sequence between residues 58 and 65 as the N-terminal sequence of the maize BEIIb that they had purified (Fisher *et al.,* 1996a,b). Moreover, they could not detect any mRNA in *ae* endosperm extracts using the *Sbe* 2b cDNA clone. Some BE activity that chromatographed as BEIIa was detected in the *ae* extracts. Although the results of Fisher *et al.* are suggestive, it remains to be shown that the activity they labeled as BEIIa in the *ae* mutant is the same as what has been termed as BEIIa in wild-type maize. It is possible that the residual enzyme activity seen in the *ae* mutant is another BE isozyme, additional to the BEII isozymes studied previously, especially because the maize lines Fisher and his collaborators used are of different genetic background from those used in the classic studies.

VI. OTHER SPECIES

In potato tubers, Vos-Scheperkeuter *et al.* (1989) purified a single form of branching activity of molecular mass 79,000. Antibodies to the native potato enzyme were prepared and the enzyme was found to react strongly only with maize BEI and very weakly with BEIIb. In neutralization tests, the enzyme inhibited the activities of both the potato tuber BE and of maize BEI. It was concluded that the potato BE shows a high degree of similarity to the maize BEI and, to a lesser extent, to the other maize BE.

However, whether potato tubers have two isoforms of BE had not been resolved. Borovsky *et al.* (1975) isolated from potato tubers a BE of molecular mass 85,000. This is close to the mass of 79,000 found by Vos-Scheperkeuter *et al.* (1989). It has been claimed that a BE of molecular mass 97,000 and 103,000 can be isolated (Blennow and Johansson, 1991; Khoshnoodi *et al.,* 1993). It is still not resolved whether the previous lower molecular mass values of 79,000 and 85,000 are the results of proteolysis during purification of the 103,000 BE or whether these proteins could be the products of different allelic forms of the BE gene or different BE genes. It is of interest to note that BEs isolated from other plants, bacteria, or mammals have molecular masses ranging from 75,000 to approximately 85,000. These molecular masses have been consistent with the molecular weights obtained from deduced amino-acid sequences obtained from isolated genes or cDNA clones.

In potato and cassava, there is no clear evidence yet for the existence of more than one isoform. Antisense experiments in potato tuber did not eliminate BE activity completely. If there were two genes coding for enzymes, antisense neutralization if one of them would not affect the expression of the other. However, this result is inconclusive because antisense experiments usually decrease the amounts of the protein encoded but do not reduce it to zero. Purification and characterization of the remaining activity in antisense experiments could resolve the matter, but was not done in this particular case.

Smyth (1988), using Indica rice, resolved two or three peaks of BE on an anion-exchange chromatography. Although the molecular weight of the BE seemed to be only about 40,000, activity was determined using the iodine assay and a crude extract, and amylase contamination or possible proteolysis apparently were not taken into account.

Mizuno *et al.* (1992) has reported four forms of BE from immature rice seeds that were separated by chromatography on DEAE-cellulose chromatography. It seems that two of the forms, BE1 and BE2 (composed of BE2a and BE2b) were the major forms, whereas BE3 and BE4 were minor forms comprising less than 10% of the total BE activity. The molecular weight of the BEs were: BE1, 82,000; BE2a, 85,000; BE2b, 82,000; BE3, 87,000; BE4a, 93,000; and BE4b, 83,000. However, BE1, 2a, and b seem to be immunologically similar in their reaction to maize endosperm BEI antibody. Moreover, the rice seed BE1, BE2a, and BE2b had similar *N*-terminal amino acid sequences. All three BEs had two N-terminal sequences, TMVXVVEEVDHLPIT and VXVVEEVDHLPITDL. The latter sequence is similar to the first sequence, lacking just the first two N-terminal amino acids. Thus, although these activities came out in separate fractions from the DEAE-cellulose column, they seem to be the same protein on the basis of immunology and N-terminal sequences. BE2a, however, was 3 kDa larger. The antibody against BE3 reacted strongly against BE3 but not toward BE1 and 2a, 2b. Thus, rice endosperm, as noted for maize endosperm, had essentially two different isoforms of BE.

Because of the many isoforms existing for the rice seed branching enzymes. Yamanouchi and Nakamura (1992) studied and compared the BEs from rice endosperm, leaf blade, leaf sheath, culm, and root. The BE activity could be resolved into two fractions, BE1 and BE2, and both fractions were found in all tissues studied in different ratios of activity. The specific activity of the endosperm activity, either on the basis of fresh weight or protein, was 100- to 1000-fold greater than other tissues studied. On native gel electrophoresis, rice endosperm BE2 could be resolved into two fractions: BE2a and BE2b. Of interest was the fact that for electrophoresis of the other tissue BE2 forms, only BE2b was found. BE2a was only detected in the endosperm tissue. It appears that in rice there could be tissue-specific isoforms of BE.

Fisher *et al.* (1996b) were able to isolate two cDNAs from an *Arabidopsis thaliana* hypocotyl library that seemed to be those coding for BEIIa and BEIIb. The two cDNAs had diverged 5′ and 3′ ends, but the amino-acid sequences encoded by them were 90% identical. The two cDNAs hybridized to transcripts that showed similar expression patterns in the vegetative and reproductive tissues.

A pea variety with wrinkled seeds has a reduced starch level, about 66–75% of that seen in the round seed; also, the amylose content is about 33% in the round form, but is much higher (60–70%) in the wrinkled pea seed. Edwards *et al.* (1988) measured the activities of several enzymes involved in starch metabolism in wrinkled peas at four different developmental stages and found that BE activity was, at its highest, only 14% of that seen for the round seed. The activities of other starch biosynthetic enzymes and phosphorylase were similar in the wrinkled and round seeds. These results were confirmed by Smith (1988), who also showed that the *r*(*rugosus*) lesion (as found in the wrinkled pea of genotype *rr*) was associated with the absence of one isoform of branching enzyme. Edwards *et al.* (1988) proposed that the reduction in starch content observed in the mutant seeds is caused indirectly by the reduction in BE activity through an effect on the starch synthase, suggesting that, in the absence of branching enzyme activity, the starch synthase activity is not optimal in α-1→4-glucan synthesis. Indeed, in a study of rabbit muscle glycogen synthase (Carter and Smith, 1978) it was found that prolonged elongation of the outer chains of glycogen caused it to become an ineffective primer, thus decreasing the apparent affinity of the glycogen synthase for the primer. It was also found that the concentration of ADPglucose was higher in the wrinkled pea than in the normal pea, suggesting that activity of the starch synthase was restricted in vivo. Under optimal in vitro conditions, in which a suitable primer such as amylopectin or glycogen is added, starch synthase activity in the wrinkled pea was equivalent to that found in the wild type.

The *r* locus of the pea seed was cloned using an antibody toward one of the pea-branching enzyme isoforms and screening a cDNA library (Bhattacharyya *et al.*, 1990). It appears that the branching enzyme gene in the wrinkled pea contains an 800-bp insertion, causing it to express an inactive branching enzyme. The sequence of the 2.7-kb clone showed more than a 50% homology to the glycogen-branching enzyme of *E. coli* (Baecker *et al.*, 1986), and the authors concluded that the cDNA that they had cloned corresponded to the starch-branching enzyme gene of the pea seed.

VII. RELATIONSHIP BETWEEN STRUCTURE AND FUNCTION

After purifying and characterizing the enzymes and cloning the genes, the next step is to understand the relationship between the structure of the branching enzymes and catalysis—in other words, the way in which variations in protein structure affect the substrate preference, the length of the branches formed, and the pattern of branching. This knowledge would obviously facilitate the manipulation of the branching enzyme gene to obtain the desired change in enzyme function. Subsequent transformation of the altered gene into plants would result in starch granules with altered structure and physical properties that could be used by the starch industry with positive economic effects.

In addition to the studies discussed in the preceding (e.g., the size of the chains transferred by the maize endosperm BE to form the α-1,6 linkages), no studies have been conducted with respect to identifying amino-acid residues involved in catalysis. However, amino-acid sequences deduced for the branching enzymes from several bacterial genes, and for many plant BE isozymes (from the cDNA clones), show high identity, leading to some information on possible amino-acid sequences that may be important for catalysis. It is of interest that the glycogen-branching enzyme from the photosynthetic cyanobacterium, *Synechococcus* sp. PCC7942, has been cloned (Kiel *et al.*, 1989) by using the *glgB* gene from *E. coli* (Baecker *et al.*, 1986) as a hybridization probe. The sequence for the cyanobacterial *glgB* gene was determined (Kiel *et al.*, 1990) and its deduced amino-acid sequence has extensive similarity to the amino-acid sequence (62% identical amino acids) in the middle area of the *E. coli* protein. It appears, therefore, that branching enzymes in nature have extensive homology, whether their specificity is for high branching, as observed in glycogen (~10% α-1,6 linkages), or a lower degree of branching, as that seen in amylopectin (~5% α-1,6 linkages).

Sequence comparisons done by a number of groups, most notably by Svensson and her colleagues (Svensson, 1994), indicate that the various starch- and glycogen-branching enzymes contain consensus sequences to the four regions that are postulated to be the catalytic regions of the α-

amylase family of enzymes. This family includes pullulanase, isoamylase, glucosyl transferase, and cyclodextrin glucanotransferase (Fig. 6). The four regions are in the central portion of the amino-acid sequences of these enzymes.

The conservation of the putative catalytic sites of the α-amylase family in the starch- and glycogen-branching enzymes should be no surprise, as BE catalyzes two consecutive reactions for the synthesis of α-1,6 glucosidic linkages: cleavage of α-1,4 glucosidic linkages and then transfer to C-6 hydroxyl groups of a glucose residue in the growing polysaccharide. These reactions are probably similar to those catalyzed by the other α-amylase family enzymes, namely cleavage and transfer to another glucose residue or to water.

Some of the eight highly conserved amino-acid residues of the α-amylase family may also be functional in branching enzyme catalysis. Research using

	Region 1	Region 2
B. subtilis α-amylase	97 **D**AVIN**H**	171 GF**R**F**D**AAKH
B. sphaericus cyclodextrinase	238 **D**AVFN**H**	323 GW**R**L**D**VANE
P. amyloderamosa isoamylase	291 **D**VVYN**H**	370 GF**R**F**D**LASV
K. pneumonia pullulanase	602 **D**VVYN**H**	673 GF**R**F**D**LMGY
Maize endosperm BE I	277 **D**VVHS**H**	347 GF**R**F**D**GVTS
Maize endosperm BE II	315 **D**VVHS**H**	382 GF**R**F**D**GVTS
Potato tuber BE	355 **D**VVHS**H**	424 GF**R**F**D**GITS
Rice seed BE 1	271 **D**VVHS**H**	341 GF**R**F**D**GVTS
Rice seed BE 3	337 **D**VVHS**H**	404 GF**R**F**D**GVTS
E. coli glycogen BE	335 **D**WVPG**H**	400 AL**R**V**D**AVAS

	Region 3	Region 4
B. subtilis α-amylase	204 FQYG**E**ILQ	261 LVTWVES**HD**
B. sphaericus cyclodextrinase	350 IIVG**E**VWH	414 SFNLLGS**HD**
P. amyloderamosa isoamylase	412 RILR**E**FTV	499 SINFIDV**HD**
K. pneumonia pullulanase	702 YFFG**E**GWD	826 VVNYVSK**HD**
Maize endosperm BE I	402 TVVA**E**DVS	470 CIAYAES**HD**
Maize endosperm BE II	437 VTIG**E**DVS	501 CVTYAES**HD**
Potato tuber BE	453 VTMA**E**EST	545 CVTYAES**HD**
Rice seed BE 1	396 TIVA**E**DVS	461 CVTYAES**HD**
Rice seed BE 3	459 ITIG**E**DVS	524 CVTYAES**HD**
E. coli glycogen BE	453 VTMA**E**EST	517 NVFLPLN**HD**

FIG. 6. Primary structures of several branching enzymes compared with the four conserved regions of the α-amylase family. The sequences have been derived from references cited in this text and in Svensson (1994). Four types of enzymes from the amylase family are compared with the branching enzymes. Svensson (1994) compared more than 40 enzymes ranging from amylases, glucosidases, several α-1,6-debranching enzymes, and four branching enzymes. The conserved amino-acid residues are in bold letters.

site-directed mutagenesis (Kuriki *et al.,* 1996) suggests that the conserved Asp residues of regions II and IV and the Glu residue of region III are important for BE activity. Their exact functions are unknown, and further experiments, such as chemical modification and analysis of the three-dimensional structure of BE, are required for determination of their role as catalytic residues and for detailing the mechanism of reaction. Other studies with phenylglyoxal, a reagent for modification of Arg residues, have shown that the maize endosperm BEs can be inactivated (Cao and Preiss, 1996). The phenylglyoxal inactivation can be prevented by the addition of amylose to the modification reaction mixture. Thus, arginine residues in BE may be involved in the binding of the substrate or in catalysis. Similarly, diethyl pyrocarbonate, a reagent specific for histidine residues, also inactivates maize BE activity, and this inactivation can also be prevented by the presence of amylose (Funane and Preiss, unpublished experiments, 1996).

It should also be noted that the amino-acid sequences at the N- and C-termini of the various BEs are dissimilar; these regions may be important with respect to substrate specificity as well as to the size of chain transferred and to the extent of branching.

Kuriki *et al.* (1997) described the construction of chimeric enzymes made from the maize branching-enzymes. The maize branching-enzymes differ in specificity of substrate and pattern of branching of the substrate. Comparing the amino-acid sequences BEI and mBEII of maize, the identity is 58%, with the identity higher (67%) in the central portion of the enzymes, which contains the regions I–IV highly conserved in the α-amylase family (Baba *et al.,* 1991; Takata *et al.,* 1992; Jespersen *et al.,* 1993; Guan *et al.,* 1994b). When amino-acid residues with similar functional side chains are taken into consideration, the two enzymes are 75% similar, with similarity 94% for the central region. Conversely, the amino and carboxy-terminal sides are quite dissimilar, suggesting that these structural differences may be responsible for the functional differences between the isozimes. To test this hypothesis, several different chimeric enzymes were constructed (Fig. 7).

Several of the chimeric enzymes constructed and expressed in *E. coli* were inactive and some had little activity, but the purified mBEII-I BspH1, in which the carboxy-terminal part of mBEII was exchanged for that of mBEI at the *Bsp*H1 restriction site (Fig. 7) was active. The resulting enzyme had properties different from both mBEI and mBEII (Table III). Activity was higher when assayed with the phosphorylase stimulation assay. In its preference for amylose (rather than amylopectin), the chimera was similar to mBEI, suggesting that the carboxy-terminal end is the region involved in determining substrate specificity and catalytic capacity. Conversely, the chimeric enzyme transferred shorter chains (d.p. around 6) than mBEI, mimicking mBEII and suggesting that the amino terminal of maize

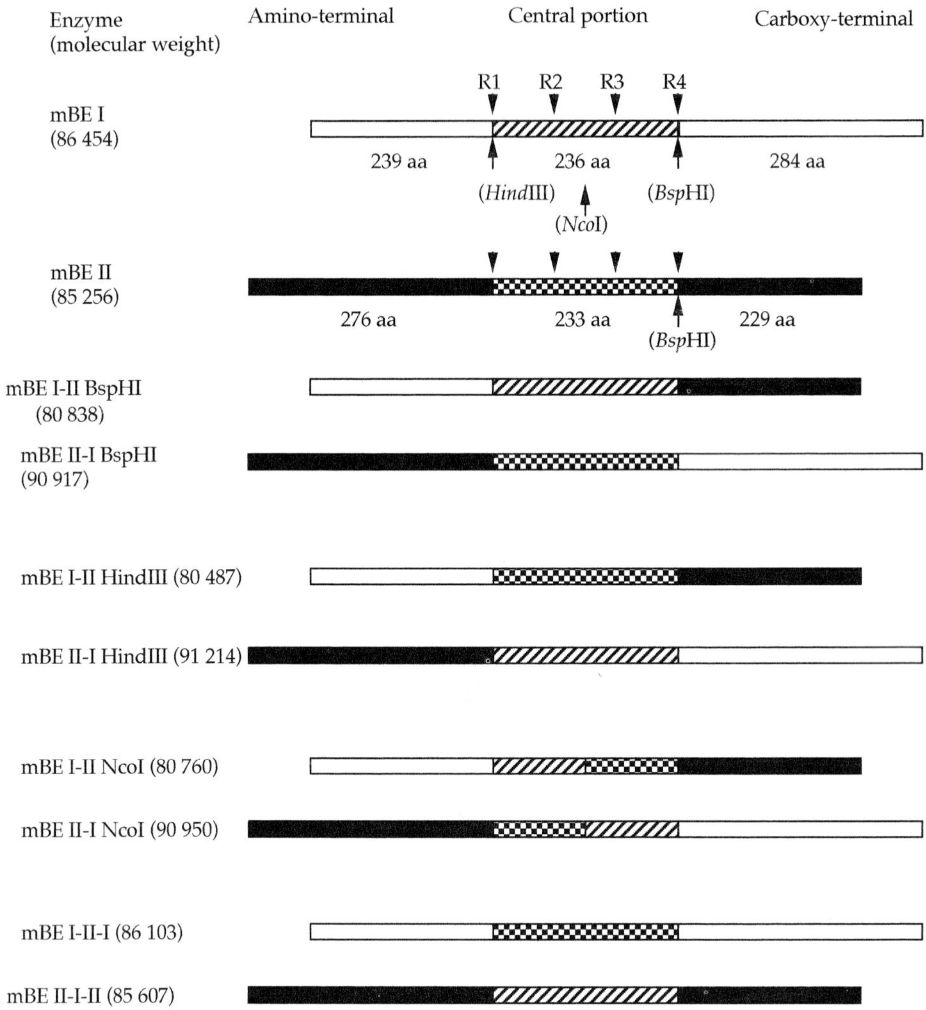

FIG. 7. Schematic diagram representing the structure of the branching enzymes from wild-type maize, mBE I, and mBE II, and of the chimeric enzymes built by genetic manipulation. The figures show amino- and carboxy-terminuses and the four conserved regions (R1 to R4) in the central portion. Some of the restriction sites used to build the chimeras were already present in the cDNAs of the branching enzymes (e.g. *Hind*III, *Nco*I and *Bsp*HI), but one restriction site had to be introduced in the sequence of mBE I by site-directed mutagenesis. The mutation introduced with the purpose of facilitating genetic manipulation has no effect on the amino-acid sequence of the wild-type enzyme and is called a "silent" mutation. The portions of the chimeric enzyme from the N- and/or C-terminal of mBE I are shown in white, whereas those from mBE II are shown in black. The central portion from mBE I is striped and from mBE II has squares. Molecular weights are in Daltons.

TABLE III

SPECIFIC ACTIVITIES OF THE TWO MAIZE ENDOSPERM BRANCHING-ENZYME ISOFORMS, mBEI AND mBEII AS COMPARED TO THAT OF THE CHIMERIC ENZYME mBE II-I BspHI[a]

Branching-enzyme	mBEI	mBEII	mBE II-I BspHI
Assay a, phosphorylase stimulation	1196	1040	3880
Assay b, branching linkage			
AS320 (100 μM)	2.1	0.4	1.3
AS110 (100 μM)	1.3	0.2	0.89
AS70 (100 μM)	0.32	0.03	0.48
Assay c, iodine stain			
Amylose (c_1)	90	6.4	69
Amylopectin (c_2)	2.3	97	2.3
Ratio of activity (c_1/c_2)	40	0.066	30

[a] Activity is expressed as U/mg of protein. For details on how the chimeric enzyme was created and how the different types of assays were performed, please see text.

branching-enzymes plays an important role in the size of the oligosaccharide chain transferred.

In all branching enzymes, but not in other amylolytic enzymes, an acidic amino acid such as Glu or Asp follows region three. In *E. coli* this acidic amino acid is Glu-459, which follows the amino acids that form part of the catalytic site and is located in a region predicted as a β-strand–α-helix loop known to host the reaction center of α/β-barrels. However, although the conservation of an amino acid throughout evolution suggests that the amino acid may be significant for the activity of the enzyme, only site-directed mutagenesis can corroborate such a hypothesis. Binderup and Preiss (in press) constructed the mutants E459A, E459D, E459K, and E459Q of the *E. coli* branching enzyme, expressed the mutated genes, and purified the mutant enzymes. The purified enzymes were then characterized using the different activity assays available for branching enzyme. Mutation of Glu-459 did not result in loss of activity. Some activity remained even when Glu-459 was substituted with an amino acid of opposite charge (E459K). The similar responses of activity to changes in pH by the wild type and the E459A mutant also ruled out the possibility that Glu-459 may be involved in acid–base catalysis. Binderup and Preiss (in press) also concluded that Glu-459 is unlikely to be involved in chain transfer, as the HPAEC-PAD chromatograms were practically identical for all constructed mutants.

A three-dimensional structure is required before a specific role can be assigned to Glu-459 and to understand why this residue is a conserved acidic amino acid in ali known branching-enzymes. Although Binderup (1997) has obtained crystals of *E. coli* branching enzymes, an essential step in constructing a reliable three-dimensional model of the enzyme, the crystals are of the needle cluster kind, unsuitable for X-ray diffraction studies. Because the only possible approach in obtaining suitable enzyme crystal is trial and error, it is impossible to predict when (and if) a three-dimensional model will be obtained.

It is worth noting that a conservative change from Glu to Asp resulted in higher specific activities in all three types of assays. It is interesting that all higher plant branching-enzymes have an Asp at the position occupied by Glu-459 in *E. coli.*

The branching enzyme from *Bacillus stearomophilus* decreased the molecular size of synthetic amylose. On studying the product of this reaction, it was found that BE had catalyzed the intramolecular transglycosylation to form a cyclic structure with a side chain. After removing the cyclic part of the molecule (using isoamylase) from the rest of the molecule, its cyclic nature was confirmed by the use of mass spectrometry. The authors proposed a new mechanism for the action of BE and suggested that plant BE may catalyze the cyclization of amylose and amylopectin.

The understanding of how branching enzymes work is progressing quickly, and this progress is likely to lead to applications to biotechnology in the near future, with the transformation of altered branching enzyme genes into crop plants resulting in starch with an altered structure and physical properties that are advantageous to the starch industry.

OPEN QUESTIONS AND HYPOTHESES IN STARCH BIOSYNTHESIS

What roles do the starch synthase isoforms play in the formation of the crystalline starch granule and amylopectin structures? How is amylose formed? Why are starch granules from different species different in size and in the number per cell? New methodology and much effort have resulted in major advances in the understanding of starch biosynthesis, but many questions remain unanswered. Here we discuss some of these open questions and possible answers.

I. INITIATION OF STARCH BIOSYNTHESIS

The synthesis of starch as described in the chapter, "The Biosynthetic Reactions of Starch Synthesis," requires the presence of a glucan primer, that is, a glucan chain that is then elongated by the addition of a glucosyl group in the following reaction:

$$\text{ADP-glucose} + [\text{glucosyl}]_n \rightarrow \text{ADP} + [\text{glucosyl}]_{n+1}$$

This equation, as written, poses the question of how the primer glucosyl_n is first formed. It seems, however, that many enzymes capable of elongating glucans can display activity independent of added primer when assayed under suitable conditions, especially in the presence of citrate. Whether these unprimed reactions proceed in vivo is unknown. Citrate has been shown to decrease, by a large factor, the K_m for primer of several enzymes: the chloroplastic starch phosphorylase (Sivak, 1992), and some starch (Boyer and Preiss, 1979; Pollock and Preiss, 1980) and glycogen (Fox *et al.*, 1976; Holmes and Preiss, 1979) synthases. Because enzyme preparations often contain some glucan brought from the material from which they were isolated, activity independent of added primer is unlikely to represent de novo synthesis of carbohydrates in most of the systems examined. In the case of the glycogen synthase from *Escherichia coli,* which was studied in great detail, the glucose is incorporated into minute amounts of glucan primer that is associated (Fox *et al.*, 1976), but is not covalently bound

(Holmes and Preiss, 1979) with the enzyme. A similar situation is that of the chloroplastic phosphorylase of spinach leaf (Sivak, 1992).

Common criteria for the identification of an unprimed product as a proteoglucan have been tested and are discussed elsewhere (see, e.g., Sivak, 1992)—that is, precipitation with trichloroacetic acid (TCA), hydrolysis with dilute acid, and extensive treatment with proteases—and they have been shown to be unsuitable for the identification of an unprimed product of phosphorylases or glycogen or starch synthases as proteoglucans. A property of many amyloses, and of starches in general, can lead to confusion and artifacts. During electrophoresis, a small part of the polysaccharide migrates into the polyacrylamide gel in the presence of the sodium dodecyl sulfate and the urea used to break noncovalent bonds. The presence of the polysaccharide can be revealed by using an iodine reagent, with the color depending on the degree of ramification of the polyglucose chain. This peculiar property of polysaccharides, which mimics protein behavior, has been attributed to the capacity of these glucans to form complexes with the negatively charged sodium dodecyl sulfate (SDS) (as they do with iodine). Also, the products of phosphorylase and of glycogen and starch synthase independent of added primer are long, mostly linear α-1,4-glucans, which because of their linearity, tend to precipitate in aqueous 5% or 10% TCA, a property that can mislead researchers into believing them to be glucosylated proteins.

Is the reaction independent of added primer relevant? This question cannot be answered, but mention can be made of some matters that must be resolved before it can be answered. The first is whether in vivo glycogen and starch synthases are able to use ADPGlc in the absence of glucan (i.e., whether the conditions are favorable for the reaction). The second is whether the site of starch synthesis, the amyloplast, is ever completely devoid of glucans that could act as primers for the starch synthase.

Speculation notwithstanding, enzymes have been isolated from potato tuber and maize endosperm, which catalyze the formation of a primer suitable for starch synthases. Studies using a particulate (sedimentable at high speed, membrane-containing) fraction of potato tuber provided evidence for the synthesis of α-1,4-glucosidic chains covalently bound to protein. The glucosyl donors in these reactions were UDPglucose, ADPglucose, and glucose-1-P (Lavintman and Cardini, 1973; Lavintman *et al.*, 1974). On the basis of other studies (Tandecarz and Cardini, 1978, 1979), it was proposed that, at least in potato tuber, a two-step reaction occurred involving a protein glucosyl acceptor, as follows:

1. UDPglucose + acceptor protein → acceptor protein-glucose + UDP
2. acceptor protein-glucose + nADP(UDP)glucose/glucose 1-P → acceptor protein-glucose-(glc)n + nADP(UDP) or nP_i

The first reaction would be catalyzed by a UDPglucose : protein transglucosylase (UPTG), and the second reaction by either a starch synthase or a phosphorylase. Moreno *et al.* (1986) solubilized and partially purified the components that catalyzed reaction 1. Ardila and Tandecarz (1992) and Bocca *et al.* (1997) purified UPTG to electrophoretic homogeneity. It is not certain whether the transglycosylase and the acceptor protein are different or the same molecule. If they are one, then the transglucosylase must self-glucosylate, as suggested by the fact that an apparently pure fraction conserved catalytic activity. The acceptor protein was determined to have a molecular mass of 38,000, and only one glucose moiety was transferred to the protein. A β-elimination reaction carried out in the presence of a reducing agent showed that an *O*-glucosidic linkage was formed and that the amino acids Ser and Thr were involved; the reaction required $MnCl_2$. Specific phosphorylases and starch synthases in the potato tuber were able to use the product of reaction 1, the glucosylated acceptor protein, as a primer to synthesize α-1,4-glucan chains (Moreno *et al.,* 1987).

The reactions involving the UDPglucose : protein transglucosylase are similar to the reactions proposed for the initiation of glycogen synthesis in mammals (Pitcher *et al.,* 1987, 1988; Lomako *et al.,* 1988). Glycogenin, a 37-kDa protein, in association with glycogen synthase, self-glucosylates to form a glucosyl protein that can act as a primer for the glycogen synthase. The glycogenin protein has been sequenced (Campbell and Cohen, 1989) and, in contrast to the plant acceptor protein, the glucosidic linkage formed is with a tyrosine hydroxyl group (Smyth, 1988). Krisman (1972) was the first to postulate that the de novo synthesis of glycogen required an initiating protein factor.

The results obtained by Cardini, Tandecarz, and their collaborators in the plant system are exciting, but some important questions remain. As we know, starch synthesis occurs in plastids (i.e., chloroplast or amyloplast). Where in the cell is the transglucosylase/acceptor protein located? If it is in the plastid, then the glucosyl donor, UDPglucose, is not available for glucosylation because it is made in the cytosol and does not enter readily the chloroplast or amyloplast. It is possible, however, that the glucosylation may occur in the cytosol and then the glucosylated acceptor protein translocates into the plastid. It is also important to note that thus far these reactions have only been demonstrated in potato tuber and maize endosperm and not in other species or organs. When cDNA coding for this protein is cloned (antibodies against the protein have been raised; J. Tandecarz, personal communication), antisense experiments will be possible. Antisense RNA is a transcript that has a high degree of complementation with a target mRNA so that it can hybridize with the mRNA in vivo. In this way, the antisense RNA acts as a repressor of the function of the target RNA. These experiments, however, are difficult to interpret because it is virtually impossible to obtain zero expression and because often plant metabolism accommodates by using an alternative pathway. If the 38-kDa protein associated with UPTG is also present in *Chlamydomonas reinhardtii* then the next step would be to find mutants lacking the 38-kDa protein and see whether starch is still synthesized in the mutant. The use of gene disruption in Antirrhynum nafus is also an attractive possibility (see Experimental Systems in the study of starch metabolism).

II. HOW IS THE STARCH GRANULE FORMED?

The glycogen-synthesizing enzymes from *E. coli* are not much different from those in plants, but bacteria make glycogen and not starch. When the maize-branching enzymes are expressed in *E. coli,* they conserve the properties they have in the plant, but the resulting glucan is glycogen; not starch. How is the intrincate structure of the starch granule formed in vivo?

The sugary mutants of maize have been known since the beginning of the twentieth century. The mutant accumulates about 35% of its dry weight as phytoglycogen, a highly branched, water-soluble polysaccharide. Phytoglycogen has 7 to 10% of its glucosidic linkages as α-1 $\rightarrow$ 6, and is therefore more highly branched than amylopectin (Manners, 1985). Pan and Nelson (1984) found that all the *su* 1 mutants were deficient in a particular endosperm-debranching enzyme activity, pullulanase, suggesting that the debranching activity is the biochemical deficiency leading to phytoglycogen formation. The debranching enzymes of normal maize endosperm were separated into three peaks of activity on a hydroxyapatite column and it was found that the *su* 1 mutant lacked one of the activity peaks toward pullulan, whereas the other two peaks were also much reduced in activity. The debranching enzyme activity of developing endosperms is proportional to the number of copies of the *Su* 1 gene, suggesting that the *Su* 1 gene is the structural gene for the debranching enzyme. The debranching enzymes, however, have not been characterized to a great extent (but see Lee *et al.,* 1971).

The observation that debranching enzyme deficiency was associated with the presence of phytoglycogen in the *su* 1 mutants (Pan and Nelson, 1984) revived a hypothesis of Erlander (1958), who proposed that amylopectin synthesis was due to debranching of phytoglycogen, which was first formed via starch synthase and branching enzyme catalysis. Although phytoglycogen may not be a normal intermediate in the synthesis of amylopectin, this may occur from a more highly branched α-1 $\rightarrow$ 4-glucan that is formed via the action of the starch synthase and branching enzyme isoforms. If the activity of the debranching enzyme is insufficient, then a more highly branched, water-soluble glucan could accumulate with a concomitant decrease in the amylopectin component of the starch granule and, possibly, an increase in the relative amylose content of the granule.

The hypothesis that the *sugary* 1 mutation affects the structural gene for a debranching enzyme is further supported by the isolation of a cDNA of the *su* 1 gene. However, its deduced amino-acid sequence is similar to a bacterial isoamylase (James *et al.*, 1995) rather than to a pullulanase.

In *C. reinhardtii,* Ball and his collaborators (Mouille *et al.*, 1996) generated seven independent alleles in the *sta*7 locus. All mutants lacked granular starch, but contained a water-soluble polysaccharide, similar to maize phytoglycogen, in an amount equivalent to 5% of the starch content of the wild type. This defect was associated with the disappearance of a specific debranching activity. All other starch-related enzyme activities were normal.

It remains to be shown whether the *su* 1 gene product debranching enzyme activity is actually an isoamylase or a pullulanase (Hizukuri, 1995). Isoamylases, such as that present in *Bacillus amyloderamosa,* readily debranch amylopectin. Pullulanase, such as that present in *Aerobacter aerogenes,* completely debranches amylopectin, but its action on glycogen is usually incomplete. The specificity of these reactions should be studied further with respect to the factors that determine which α-1,6 linkages are cleaved and which remain resistant to debranching action. It is possible that the crowding of the α-1,6 linkages in a cluster region in amylopectin causes some steric difficulties for the debranching of the linkages in the cluster region, but at present this is only conjecture. As for the problem of starch initiation, this is a research field that would benefit greatly from gene disruption experiments.

III. A COMPLETE PATHWAY

From the information available on the branching enzyme and starch synthase isozymes, a possible route for the synthesis of amylopectin and amylose can be proposed as shown in Fig. 1.

A reaction with the potential of being the initiating reaction for synthesis has been observed in potato tuber and maize endosperm (see preceding). The resulting glucosylated 38-kDa protein can serve as a primer for the synthesis of starch via the starch synthase reactions. Whether there is an acceptor protein that could be glucosylated by ADPGlc has not been demonstrated; the proposed initiating reaction and acceptor protein have not been characterized as well as the other reactions in starch synthesis.

After the formation of the unbranched maltosaccharide–protein primer of undetermined size, high rates of polysaccharide formation may occur at

1. Initiation of synthesis of unbranched maltodextrins (bound to protein?)

GBSS + SSSII + BEI
2. Unbranched maltodextrin + ADPGlc ---------------> Long, intermediate size chains of glucan

BEII + SSS I
3. Long, intermediate size chain glucan + ADPGlc -----------------> Synthesis of A & shorter B Chains to finish cluster Structure

4. Repeat of reactions 2 and 3 to form the complete pre-amylopectin polysaccharide.

Debranching Enzyme
5. pre-amylopectin structure ------------------> amylopectin + pre-amylose chains

GBSS
6. pre-amylose chains --------------> amylose

FIG. 1. Hypothetical pathway for the synthesis of amylose and amylopectin. Initiation may involve synthesis of a maltodextrin attached covalently to a protein. This putative protein-α-glucan then can accept glucose from ADPGlc, either via GBSS catalysis to form an amylose structure, or in combination with BEI, SSSII, and (possibly) GBSS to form a polysaccharide having the internal structure of the final amylopectin product. BEII and SSSI carry out the reactions to form the exterior of the amylopectin structure. The enlargement of the amylopectin could proceed further by continuing participation of BEI, SSSII, and (possibly) GBSS, by repeat of reactions 2 and 3. Production of amylopectin in reaction 5 is caused by debranching enzyme, which also generates oligosaccharide chains, which are elongated by GBSS to form the amylose fraction.

the surface of the developing starch granule, where granule-bound starch synthase (GBSS), soluble starch synthase II (SSSII), and branching enzyme I (BEI) interact with the glucosylated protein primer to form a branched α-glucan containing both long and intermediate-size chains.

The postulation of phase 2 in Fig. 1 is based on the studies of the polysaccharide structures observed in the *C. reinhardtii* mutants deficient in SSSII and GBSS (Fontaine *et al.,* 1993; Maddelein *et al.,* 1994a,b), as well as the *ae* mutants of rice (Mizuno *et al.,* 1993) and maize (Boyer and Preiss, 1981), which are defective in branching enzyme II (BEII). BEII deficient mutants have altered polysaccharides with fewer branches and longer sized branched chains. In phase 3, SSSI and BEII are responsible for the synthesis of the A- and exterior B-chains to complete the first cluster region in the glucan. Continued synthesis in phase 4 is essentially a repeat of phases 2 and 3 to synthesize a highly branched α-glucan, termed *proamylopectin.* This highly α-branched glucan is water soluble and noncrystalline. In phase 5, a debranching enzyme debranches the preamylopectin to form amylopectin, which can now crystallize. In phase 6, the chains, liberated by debranching action of the proamylopectin could be used as primers by GBSS to form amylose. Amylose synthesis may occur only inside the starch granule, and only GBSS would be involved because it may be the only starch synthase present at the site of amylose synthesis.

These reactions do not have to occur in perfect sequence, and the phases may have some overlap (e.g., phases 2, 3, and 4 may overlap, and possibly even 5 and 6). However, the present evidence, such as intermediate products formed by starch mutants of *C. reinhardtii* and of higher plants, supports the sequence of reactions shown in Fig. 1 for amylopectin and amylose biosynthesis. Further experiments are required to test this hypothetic scheme, and attempts to purify and characterize the debranching enzyme, crucial to this hypothesis, are under way.

It should be noted that the proamylopectin in this still hypothetical pathway would be larger than the phytoglycogen found in the mutants lacking debranching activity. This is because proamylopectin would have a size comparable to amylopectin, while phytoglycogen, much smaller, may be the product of degradation of a proamylopectin unable to crystallize into amylopectin and may be so unprotected that it would be subject to the action of amylases.

THE SITE OF STARCH SYNTHESIS IN NONPHOTOSYNTHETIC PLANT TISSUES: THE AMYLOPLAST

A hypothetic metabolic pathway must pass a crucial test to be accepted by the scientific community: Are the enzymes in the "right" compartment within the cell so that the product of one reaction can be used by the next enzyme in the pathway? Starch granules are large enough to be observed with a microscope (see Fig. 2 in the chapter, "Physicochemical Structure of the Starch Granule"). With good techniques, it is possible to see that the starch granule is enclosed within a defined structure within the cell. Are the ADPGlc PPase, the starch synthase, and the branching enzyme also there?

A large part of the biosynthetic capacity of a plant cell is localized in plastids, which are self-replicating organelles surrounded by a double-membrane envelope. Plastids are present in most cells of photosynthetic eukaryotes. In most angiosperms, however, sperm cells lack plastids, a fact that makes plastid inheritance solely maternal. The envelope is composed of an outer and inner membranes, which differ in their permeability, separated by a 10 to 20-nm gap. The plastids contain DNA that is concentrated in a section of the stroma, which is the background matrix of the plastid. The plastidial ribosomes are smaller that the cytoplasmic ribosomes.

The green chloroplasts are the site of photosynthesis; chromoplasts contain carotenoid pigments and can be found in flowers, fruits, senescing leaves, and sometimes in roots; etioplasts are present in seedlings grown in the dark. Storage plastids are designated according to the nature of the product they accumulate (i.e., protein in proteinoplasts, lipids in elaioplasts, and starch in amyloplasts). Plastids from different tissues can differ both in morphology and biochemistry, but the DNA they contain is identical (Dennis *et al.*, 1985). This indicates that all plastids arise from the same precursor plastids and later differentiate in response to the development of the tissue. For a proposal of how proplastids develop into the different types of plastids in a plastid cycle, see Whatley (1978).

Amyloplasts are characterized by the presence of one or more starch granules that grow in size as the storage organ develops, distending the plastid. The structure of the amyloplast can be studied using microscopy, but preparation of the samples requires great care to avoid artifacts. Microscopy can also help in the localization of enzymes within the amyloplast. The study of amyloplast biochemistry and transport demands "good" amyloplasts (i.e., functional, whole plastids, clean of contaminating enzymes belonging to other cellular fractions, and with intact envelopes).

I. MICROSCOPY AND IMMUNOCYTOCHEMICAL STUDIES

Kim *et al.* (1989) employed immunocytochemistry at the light microscopy level. An antibody raised against the spinach leaf ADPGlc PPase (which specifically reacts with the potato ADPGlc PPase in immunoblotting experiments) was used on thin sections of the potato tuber, and the bound antibody was detected with either a secondary fluorescent antibody or with protein A tagged with gold particles (Fig. 1). The ADPGlc PPase was specifically localized within the amyloplast, confirming the results obtained by others (Mohabir and John, 1988) using a different methodology (see later).

Ultrathin sections for electron microscopy (thickness approximately 100 nm) must be able to withstand the electron beam and the vacuum in the microscope. For this, it is first necessary to stabilize the ultrastructure of the fresh tissue by fixation, then to dehydrate it with an organic solvent, and finally to embed it in a resin. The resulting hard block can be cut into ultrathin sections, which are then mounted on a grid and stained. For immunogold labeling, free-aldehyde groups and nonspecific binding sites

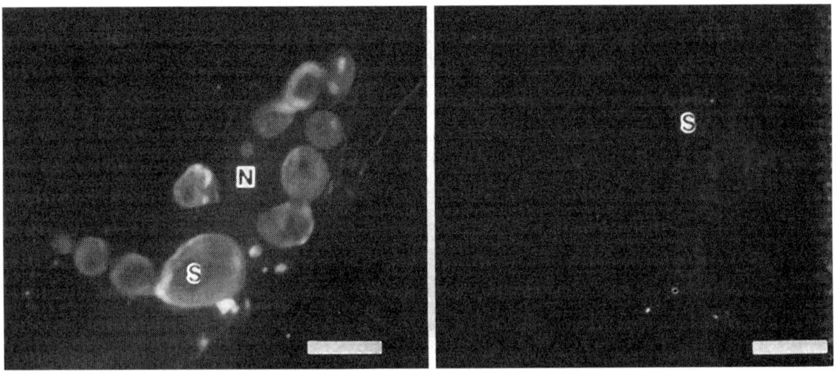

FIG. 1. *Left:* Immunofluorescence labeling for ADPGlc PPase demonstrating intense fluorescence only in the amyloplasts. Bar = 10 μm. *Right:* Preimmune control for immunofluorescence labeling. The background fluorescence is low throughout the cell. Figure reprinted with permission from Kim *et al.* (1989).

are saturated with buffers containing glycine and gelatin, and the grids are incubated with a suitable concentration of the antiserum. After washing excess serum, the sections are incubated with protein A-gold, then washed, dried, and stained with aqueous uranyl acetate (Kram, 1995).

Starch storage tissues are difficult to prepare for ultramicroscopy, because starch granules are often incompletely fixed, and as a consequence, they fold and detach from the rest of the section. Kram (1995) obtained good sections for ultrastructural research using conventional embedding of potato microtubers in Epon resin (Fig. 2). For immunolocalization, she found that slow cooling during dehydration and embedding in Lowicryl K4M (Norticon, Breda, The Netherlands) at −30°C gave the best results. Amyloplasts in potato microtubers are not identical to those present in normal tubers but are, in many respects, a good experimental system. Indeed, starch granules in potato tubers are very large, adding to the difficulty in preparing good-quality sections, but the granules in microtubers are smaller.

Electron micrographs of castor bean endosperm tissue showed the presence of proplastids with starch grains. Starch synthase was shown to be associated with the proplastid fraction of the endosperm tissue along with the starch granule (Reibach and Benedict, 1982).

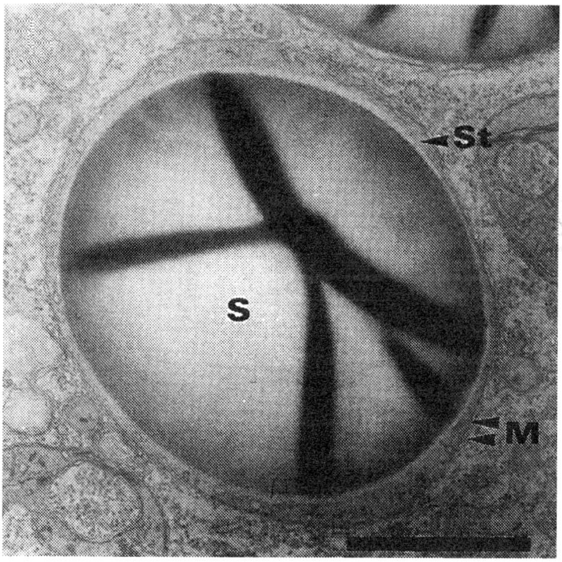

FIG. 2. Electron micrograph showing a potato microtuber amyloplast. S, starch; M, amyloplast membranes; St, amyloplast stroma. Bar = 1 μm. Figure reprinted with permission from Kram (1995).

II. CELL FRACTIONATION

Fractionation techniques relying on differential centrifugation, of the kind used for isolation of chloroplasts (see the chapter, "Starch Accumulation in Photosynthesis Cells), are inadequate because the amyloplast contains large starch granules that disrupt the integrity of the membranes even under mild centrifugal forces. Macdonald and ap Rees (1983) were successful in isolating intact amyloplasts not seriously contaminated by cytosol, from soybean cell protoplasts and showed that ADPGlc PPase and starch synthase were confined to the amyloplast. Echeverria *et al.* (1985, 1988) isolated amyloplasts from protoplasts prepared from maize endosperm harvested 14 to 17 days after pollination (DAP). Similarly, Journet and Douce (1985) isolated amyloplasts from cauliflower buds, and Macheral *et al.* (1985) and Journet *et al.* (1986) isolated them from sycamore (*Acer pseudoplatanus*) cells. Entwistle *et al.* (1988) and Entwistle and ap Rees (1988) were able to isolate amyloplasts from lysates of protoplasts obtained from the endosperm of developing grains of wheat, and Mohabir and John (1988) obtained a fraction enriched in intact potato tuber amyloplasts. The findings and conclusions reached by all these groups were similar to those of Macdonald and ap Rees (1983) (i.e., that the starch biosynthetic enzymes are primarily, if not exclusively, confined to the amyloplasts of those tissues).

In short, cell fractionation, which is composed of three steps—homogenization, fractionation, and analysis—can be an excellent way to locate an enzyme within the cell. The reader is referred to the excellent commentary by ap Rees (1995), in which the rigorous criteria to follow so that a cell fractionation provides good, reliable information is summarized. The author concludes that work done on soybean protoplasts (Macdonald and ap Rees, 1983), wheat endosperm protoplasts (Entwistle and ap Rees, 1988), wheat endosperm (Tetlow *et al.*, 1993), pea embryos (Denyer and Smith, 1988), and pea roots (Borchert *et al.*, 1993) provides further support for the view that ADPGlc PPase essentially is confined to the plastid.

In short, authors using a variety of methods and plant systems reported that in nonphotosynthetic tissues the enzymes of starch biosynthesis appear to be restricted to the amyloplast. However, two reports have proposed that a significant portion of the ADPGlc PPase activity may be present in the cytosol. Amyloplasts were isolated from wheat endosperm by Thornbjørnsen *et al.* (1996), with intactness ranging from 41 to 89%. The proportion of enzymatic activity recovered in the amyloplast fraction, in relation to total activity, was 13 to 17% for starch synthase and alkaline pyrophosphatase, and only 2.5% for the ADPGlc PPase. On this basis, the authors calculated that of ADPGlc PPase activity residing in the amyloplast was 15% of the total, and that the rest was in the cytosol. Immunologic studies by the same authors detected two different isoforms of the ADPGlc PPase: one mainly cytosolic and the other mainly plastidial. The authors indicated that there is an excess of ADPGlc PPase activity in the amyloplast to account for the starch synthetic rate, and they were uncertain about the function of the putative cytosolic ADPGlc PPase.

In a report by Denyer *et al.*, 1996, preparations enriched in maize endosperm plastids contained 24 to 47% of the total activity of the plastid-marker enzymes, starch synthase and alkaline pyrophosphatase, but they contained only 3% of the total ADPGlc PPase activity. On this basis, the authors estimated that more than 95% of the ADPGlc PPase activity was nonplastidial. Using antibodies prepared against the *Bt* 2 subunit of the maize endosperm ADPGlc PPase, they showed that most of the *Bt* 2 protein was confined to the supernatant, and some was in the plastid. In *bt* 2 mutant kernels, the cytosolic protein that reacted with the *Bt* 2 antiserum was not detected, but there was a plastidial form of ADPGlc PPase. These data are somewhat different than what has been obtained by Miller and Chourey (1995) and by J. L. Prioul (personal communication, 1997) who, using immunogold labeling, detected the *Bt* 2 protein in the amyloplast. If the data from Denyer *et al.* (1996) are not artifactual, this would mean that there is more than one route for synthesis of ADPGlc in maize endosperm. Most authors believe that carbon translocated into the plastid via a glucose-6-P translocator is converted to ADPGlc by the action of the (plastidial) phosphoglucomutase and the ADPGlc PPase (Neuhaus *et al.,* 1993). Conversely, Denyer *et al.* (1996) believe that some ADPGlc synthesis goes on in the amyloplast, catalyzed via a plastidial ADPGlc PPase, but since in their model most of the ADPGlc is synthesized in the cytosol, it must be translocated into the plastid for starch synthesis; thus their model demands an ADPGlc transporter.

III. TRANSPORT OF CARBON INTO AMYLOPLASTS

As discussed previously, the model of Denyer *et al.* (1996), in which most of the ADPGlc is synthesized in the cytosol, demands an ADPGlc transporter if starch synthesis is to proceed (as it does) within the amyloplast. However, no protein with those properties has yet been identified. Although ADPGlc uptake by the *A. pseudoplatanus* amyloplasts has been reported (Pozueta-Romero *et al.,* 1991), Borchert *et al.* (1993) and Batz *et al.* (1994) showed that this ADPGlc transport may not be relevant physiologically. In vitro ADPGlc may be translocated via the ATP/ADP translocator, but since both ADP and ATP effectively inhibit ADPGlc uptake at concentrations lower than their physiologic concentrations (in pea-root and cauliflower-bud amyloplasts), in vivo transport of ADPGlc by the ATP/ADP translocator is unlikely to be relevant.

It has been suggested that the *Bt* 1 gene product may be the ADPGlc transporter. The *Bt* 1 gene encodes a plastidial membrane-associated protein (Cao *et al.*, 1995; Sullivan and Kaneko, 1995), whose deduced amino-acid sequence shows similarity to known adenine nucleotide transporters (Sullivan *et al.*, 1991). The *bt* 1 mutant is starch deficient and shows a high level of ADPGlc concentration in the endosperm as compared with the normal endosperm (Shannon *et al.*, 1996). However, this is highly speculative and the *Bt* 1 protein remains to be studied and characterized, and its function remains to be determined. In conclusion, the hypothetic ADPGlc transporter required by the model used by Denyer *et al.* (1996) remains to be found.

Hypothetic models aside studies employing a variety of methods have shown that the starch biosynthetic enzymes in leaf tissue are localized in the chloroplast (see the chapter, "Starch Accumulation in Photosynthesis Cells"), and in nonphotosynthetic tissue they are localized in the amyloplast. However, an important question remains to be answered: What metabolite is transported into the amyloplast to provide carbon and energy for starch synthesis? For the chloroplast, it is clear that the main transport system for carbon is the triose-P/P_i translocator (Heber and Heldt, 1981). It had been assumed that a similar process would also be functional in the amyloplast envelope, based on the rationale that amyloplasts could develop into chloroplasts and vice versa, and some data seemed to support this view. For example, Echeverria *et al.* (1988) isolated amyloplasts from maize endosperm able to convert labeled triose-P into starch. Mohabir and Johns (1988) also suggested that potato tuber amyloplasts have a triose-P/P_i translocator when triose-P, generated by the addition of labeled fructose-1,6-bisP with aldolase, triose-P isomerase, and fructose-2,6-bisP to an intact amyloplast fraction, was converted into starch. The plastids from cauliflower buds were shown to contain all the enzymes necessary to convert triose-P to starch (Journet and Douce, 1985). Borchert *et al.* (1989) prepared amyloplasts from pea roots and identified a translocator that exchanged P_i with glucose-6-P, dihydroxyacetone-P, or 3PGA. The translocator had low affinity for 2PGA or glucose-1-P. The highest affinity was seen with dihydroxyacetone-P and P_i, and then with 3PGA and glucose-6-P. It is possible, however, that contamination from other cellular fractions may have led to artifacts.

Other evidence, however, supports the presence of a different transport system in the amyloplast envelope. Keeling *et al.* (1988) studied starch synthesis in isolated wheat endosperm tissue or in the intact plant by incubation with [1-^{13}C]- and [6-^{13}C]-glucose and by looking at the extent of the redistribution in the glucose moieties of the starch formed. The starch was isolated and the distribution of the ^{13}C isotope was determined. If carbon flow into starch were via the triose-P isomerase, the redistribution would have been extensive. However, there was very little increase of the incidence of ^{13}C in carbons 2–5. A redistribution of 15 to 20% of label, between carbons 1 and 6 of glucose recovered, is consistent with some conversion of glucose into triose-P, resynthesis of the hexose, and its conversion into starch. Since the same redistribution was observed for sucrose, it was concluded that the redistribution occurred in the cytosol and not in the amyloplast. These data did not support a triose-P/P_i translocator as a major transport system of carbon into the amyloplast for starch synthesis, indicating that the major carbon transport system involves a sugar-P: glucose-1-P, glucose-6-P, or fructose-6-P.

Entwistle and ap Rees (1988) found that wheat endosperm lacked significant amyloplastic fructose-1,6-bisphosphatase, an enzyme that would be required if a triose-P/P_i transport system were involved in starch synthesis. In the search for a transport system to supply carbon for starch synthesis in the wheat endosperm, Tyson and ap Rees (1988) incubated intact amyloplasts with different ^{14}C-labeled compounds (i.e., glucose, glucose-1-P, glucose-6-P, fructose-6-P, fructose-1,6-bisP, dihydroxyacetone-P, and glycerol-P). Only glucose-1-P was incorporated into starch, and this incorporation was dependent on the integrity of the amyloplast. These results are consistent with the results of Keeling *et al.* (1988). Direct import of C-6 compounds has been reported for amyloplasts of potato tubers, fava beans (Viola *et al.*, 1991), maize endosperm, and suspension cells of *Chenopodium rubrum* (Hatzfeldt and Stitt, 1990). Labeled hexose monophosphates can be converted into starch by plastids isolated from wheat endosperm (Tyson and ap Rees, 1988), soybean suspension cultures (Coates and ap Rees, 1994), pea cotyledons (Hill and Smith, 1991), and cauliflower florets (Batz *et al.*, 1994).

Is there an explanation for the contradictory results obtained by different research groups? An easy explanation is the existence of very different transport systems in amyloplasts of different tissues. Another explanation is that some workers studied amyloplasts that had been damaged and/or contaminated during isolation. The permeability properties of the isolated amyloplasts depend on the degree of intactness, and the present methodology to evaluate plastid intactness measures enzyme latency (ap Rees and Entwistle, 1989) or relies on microscopic examination (Pozueta Romero *et al.*, 1991). None of these methods are sufficient to evaluate the extent of the damage suffered by the plastid envelopes or the degree of contamination of the amyloplast by other cellular components. Although intact amyloplasts are far more difficult to isolate than spinach or pea chloroplasts, it is hoped that the more recently developed methodology will be further improved, resulting in more reliable experimental results.

Although it appears that the major carbon transport system for the wheat grain amyloplast does not involve triose-P, and most likely involves hexose-P (Fig. 3), the major carbon transport systems for other amyloplasts are not certain. More recent evidence does indicate that what is true for wheat is also true for pea embryo amyloplasts, for maize, and for potato. Thus, it may be that the major transport system for most reserve, nonphotosynthetic plant systems is at the hexose-P level and not at the triose-P level. The glycolytic scheme in the amyloplast may then take on a more important function than is observed in the chloroplast, in that the scheme aids in the production of amyloplastic ATP. Amyloplastic 3PGA may then also be an indicator of the state of ATP level in the amyloplast and, on its accumula-

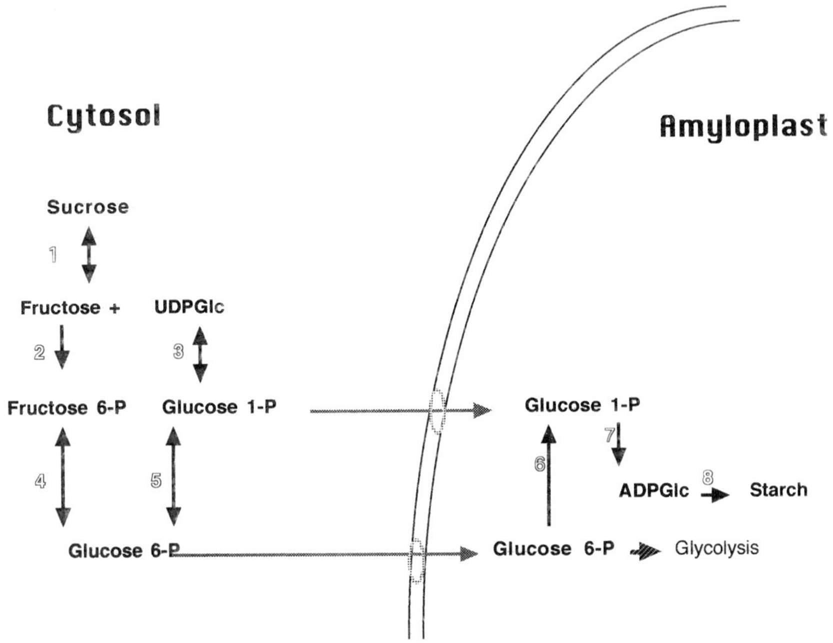

FIG. 3. Transport of carbohydrates into amyloplasts and possible routes they follow after they are inside the plastid. In the cytosol, sucrose is metabolized into glucose-6-P and glucose-1-P, which are then translocated into the amyloplast via specific translocators (ovals). The enzymatic reactions shown are as follows: 1, sucrose synthase; 2, fructokinase; 3, UDPGlc PPase; 4, cytosolic P-hexoseisomerase; 5, cytosolic P-glucomutase; 6, plastidial P-glucomutase; 7, ADPGlc PPase; and 8, starch synthase. A cytosolic ADPGlc PPase and the corresponding ADPGlc translocator proposed by some authors are not shown in this figure because the evidence supporting their existence is insufficient (see text).

tion, may be an indicator of high ATP concentration and/or carbon excess, thus stimulating starch synthesis by stimulating the ADPGlc PPase or reversing its P_i inhibition.

From the controversies that persist in the biochemistry and molecular biology of starch biosynthesis, that regarding the localization of ADPGlc PPase has been visited and revisited by researchers several times since the mid-1970s. Biochemical studies clearly indicate that the enzyme is located in the chloroplasts of leaves (see the chapter, "Starch Accumulation in Photosynthesis Cells"). Conversely (as discussed in this chapter), there is less agreement concerning the localization of the enzyme in nonphotosynthetic tissue; several methods have been used in the cereals, including immunolocalization at the electron microscopy level (Miller and Chourey, 1995) and cell fractionation followed by enzyme assay (Echeverria *et al.*, 1988); contradictory results have been published (Villand and Kleczkowski, 1994). The data obtained by Brangeon *et al.* (1997) confirm the earlier results of Miller and Chourey (1995) and provide detailed information on the correlation between the expression of ADPGlc PPase and starch accumulation within the endosperm, showing how the tissue- and cell-specific expression varies throughout the grain-filling period, as the endosperm and amyloplasts mature.

Brangeon and his collaborators (1997) studied maize kernels using light microscopy to determine the citologic structure of the fruit wall (pericarp and nucellus) and endosperm during development and filling of the grain. The authors chose four different stages representing late cellularization [8–9 days after pollination (DAP)], cell differentiation and enlargement (15 DAP and 23 DAP, respectively), and maturation of the endosperm tissue (35 DAP). For immunolocalization of the ADPGlc PPase, they used antibodies against the small or the large subunit, and they also did in situ hybridization of the corresponding mRNA transcripts.

In very young kernels, immunolabel was observed exclusively in cells of the pericarp layer, with no staining visible in the endosperm. At this early stage, there were no starch-bearing plastids in the young developing endosperm cells. By the next stage of development examined (14 DAP), the endosperm had expanded to some extent. The outer meristematic layer of cells underwent tangential divisions to produce cells to the inside, and radial divisions to extend its surface. The meristematic cells ceased to divide by 16 or 17 DAP; the outer layer cells gave raise to aleurone and subaleurone cell layers, which contained protein and lipid globules. Conversely, the endosperm contained starch, displaying a gradient of maturation in which the outer layers containing small vacuoles and storage bodies but no starch-bearing plastids; the cells toward the center were larger, with denser cytoplasm and larger starch granules within the amyloplasts. Throughout the endosperm, the strength of immunolabeling was correlated with the number and size of the starch granules. Under high magnification, a circular pattern was seen in young amyloplasts (Kim *et al.,* 1989, in potato) interpreted by Brangeon *et al.* (1997) as reflecting a higher enzyme concentration in zones active in starch synthesis.

Cytoplasmic strands were clearly immunonegative, and the immunolabel was associated exclusively with the amyloplasts. When using cDNA probes for mRNA coding for the small or large subunits of ADPGlc PPase, the label was dispersed throughout the cytosol, as expected for an enzyme encoded by nuclear genes.

By 23 DAP, cell division had ceased, as increase in endosperm volume was by cell expansion only. The pericarp had collapsed and the aleurone and subaleurone cell layers were fully differentiated. The outer layers of the starchy endosperm contained small, rounded granules embedded in cytosol, and the more central cells were filled with starch and had very little cytosol.

At 35 DAP, the endosperm had reached its maximum size and had started drying—the cells eventually dying, with starch grains filling them completely. The pericarp was crushed by expanding endosperm, becoming the final outer fruit coat composed of thick-walled dead cells.

In short, the data obtained by Brangeon *et al.* (1997) confirm the previous work of Miller and Chourey (1995) concerning the localization of the ADPGlc PPase within the amyloplasts. It is worth noting that these two groups came to similar conclusions after working with different lines, using different methodology and different antibodies against ADPGlc PPase. Conversely, the data of Denyer *et al.* (1996), proposing that most of the ADPGlc PPase in maize endosperm is associated with the cytosol, are still to be confirmed by other laboratories. Although Brangeon and collaborators (1997) suggests a number of explanations for the results of Denyer and colleagues (1996), in our view the most likely problem was the inability of Denyer *et al.* (1996) to obtain intact amyloplasts and to take into account this lack of intactness in their interpretation of the data. Until good amyloplast preparations displaying intact envelopes are obtained, immunolocalization data such as those obtained by Brangeon *et al.* (1997) and Miller and Chourey (1995) should be preferred to data such as that from Denyer and colleagues, especially in the context of all the other evidence available.

To conclude: despite proposals presented by some authors and discussed in this chapter, the experimental evidence available supports mechanisms for the transport of carbon into the amyloplast and its conversion into starch, as depicted in Fig. 3.

REGULATION OF THE STARCH SYNTHESIS PATHWAY: TARGETS FOR BIOTECHNOLOGY

I. INTRODUCTION

Glycogen synthesis in mammalian cells is relatively well understood, including the specificity of glycogen synthase for UDPglucose as well as its regulation through hormonally induced posttranslational protein modification. Textbooks of biochemistry usually describe these metabolic schemes in detail. Conversely, the biosynthesis of polysaccharides in bacteria and plants is usually described only superficially. These organisms accumulate glycogen (bacteria) or starch (plants) by metabolic pathways that are different in a number of respects from those occurring in animals. Despite the different structures of the final products, in both bacteria and plants ADPglucose is the glucose donor for the elongation of the α-1,4-glucan chain. Moreover, in both systems, the main regulatory step of the metabolism takes place at the level of ADPglucose synthesis.

II. GENETIC ENGINEERING

The gene technology developed in the past few decades can be used to test scientific hypotheses and to alter plant metabolic pathways for commercial advantage. These two uses of technology go hand in hand: unless it is understood how a pathway works, it is very difficult to change that pathway in a particular direction.

Plant transformation is now an experimental tool that can be used on many species; a large number of genes, viral genomes, and plant-transposable elements have been transferred to the genomes of species such as potato, tobacco, oilseed rape, and so on.

Transforming a plant with a foreign DNA involves a number of steps. Exogenously added DNA has to be taken up by isolated plant cells, and the transformed cells must be able to regenerate a plant; the exogenous DNA can originate in bacteria, animals, or other plants. Breeding is no longer limited to existing varieties within the species but is limited only by imagination and good biochemistry.

The methodology required for successful modification of plant metabolism has been perfected for some plant species and for specific enzymes. This methodology includes transformation of plant cells or tissues; regeneration of a healthy plant from the transformed cells; isolation of the gene encoding the enzyme of interest; identification of promoter sequences for tissue-specific expression; and targetting of the protein into the cellular compartment desired.

A transgenic organism is defined as one whose genome has been modified by the addition of exogenous DNA. This exogenous DNA can be a manipulated sequence from the same species or DNA from another species (plant or otherwise) that has a desirable property. The operational gene in the exogenous DNA is called the *transgene.* The DNA can be introduced into a cell by a variety of techniques (e.g., injection, transformation, viral infection). Biotechnology refers to the area of research in which recombinant DNA techniques are used to design and produce genotypes profitable for agriculture or other commercial enterprises.

III. VECTORS

The vectors used routinely to produce transgenic plants are derived from the soil bacterium *Agrobacterium tumefaciens.* In its natural form, this bacterium causes the crown gall disease in which the infected plant produces tumors ("galls") usually at the base ("crown") of the plant.

Part of the process of infection and tumor formation requires the insertion of the T_i plasmid of the bacterium into the genome of the plant, a characteristic that makes the T_i plasmid an ideal vector for the introduction of foreign DNA into the plant. The DNA of interest is "spliced" into the T_i plasmid, and then the whole segment is inserted into a plant chromosome. To achieve this objective, some modifications must be made to the T_i plasmid—for example, its attenuation (deletion of tumor-inducing genes), the insertion of cloning sites so that the DNA of interest can be inserted easily into the vector, and the addition of selectable genes.

The gene of interest is spliced into the modified T-DNA by conjugation of *Agrobacterium* with *Escherichia coli* containing an "intermediate vector." Then, *Agrobacterium* cells containing the recombinant plasmid are selected after conjugation by growing the culture in the presence of a suitable antibiotic, and these are the cells used to infect cut segments of plant tissues. The infected plant tissue (e.g., leaf discs) is placed in a medium containing the other antibiotic (e.g., kanamycin) so that only plant cells that acquired antibiotic resistance from the T-DNA transfer survive. The transformed cells grow into clumps of cells that can be induced to form roots and shoots when plant hormones and nutrients are present in suitable quantities in the growth medium. The plants obtained in this manner can then be screened to see whether the DNA of interest has been incorporated in their genomes. DNA that have been incorporated using this methodology include those that confer resistance to glyphosate (a herbicide) and a gene that delays ripening. Genes relevant to starch synthesis have been transformed into potato, tobacco, and tomato using *A. tumefaciens* T_i plasmid–derived vectors. *Agrobacterium* has also been used to transform seeds, and transgenic plants of *Arabidopsis thaliana* have been obtained by cocultivation of imbibed seeds with *Agrobacterium* (Feldman and Marks, 1987).

The majority of the attempts to transform plants in the 1970s failed or remained unsubstantiated. A major advance occurred when chimeric genes were constructed in which the coding regions of foreign genes were inserted between the signals controlling gene expression in plants—upstream promoters and downstream adenylation sites (Downey *et al.*, 1983). The criteria confirming the successful integration and expression of the chimeric gene in the plant cells included the phenotypic expression of the desired characteristic, Southern blots to demonstrate the presence of the DNA in transformed tissue, Northern blots to confirm the presence of the RNA transcript of the correct size, and activity of the enzyme. Later studies established the sexual transmission of the foreign DNA to progeny of the transgenic plants in segregation ratios typical of simply inherited genes.

IV. PROTOPLAST ISOLATION AND TRANSFORMATION

Not all plant species are amenable to transformation using the T_i-plasmid. In this case, protoplasts are prepared and then induced to take up exogenously applied DNA. Protoplasts can be isolated from a variety of plant tissues, although usually leaves are used. Protoplast isolation involves the enzymatic removal of the cell wall by incubating tissue slices in a medium including fungal cellulases, pectinases, and hemicellulases. The medium is prepared at a high osmotic potential to prevent the bursting of the protoplasts once the cell wall has been digested. The conditions required to prepare viable protoplasts depend on the plant species and the tissue, and must be determined empirically—that is, by trial and error. Once purified from cell debris and the enzymatic solution, the protoplasts are ready for transformation. A number of methods can be used in order to get the protoplasts to take up the exogenous DNA, including the use of polyvalent cations and electroporation, fusion of protoplasts with liposomes containing the foreign DNA, and microinjection.

If the DNA is maintained stably within the cell and if the protoplasts can be induced to regenerate a whole plant, transgenic plants can be obtained.

V. PLANT REGENERATION

A few days after isolation, the protoplasts in their culture medium begin to regenerate their cell walls and divide to form a microcallus. A *callus* is a mass of undifferentiated plant cells, and its formation is dependent on the presence in the culture medium of the plant hormones auxins and cytokinins. The callus may eventually differentiate to form shoots and/or roots, depending on the balance of plant hormones in the culture medium, as demonstrated by Skoog and Miller (1957) with tobacco callus. When the ratio of auxins to cytokinins in the culture medium was high, the calluses were induced to form roots. Conversely, a low ratio of auxins to citokinins induced the formation of shoots. Intermediate ratios promoted the growth as callus.

For some plant species it has not yet been possible to regenerate plants from protoplasts, and in these cases protocols have been designed to transform embryogenic explants rather than protoplasts. When using embryogenic explants, it may be necessary for the foreign DNA to travel through several layers before reaching the cells that will originate the germ line, and for this reason the DNA coating very small particles or tungsten or gold is delivered into the target cells using an explosive force. This approach has been used successfully on maize and soybean.

VI. TISSUE- AND ORGANELLE-SPECIFIC EXPRESSION

A promoter used for many applications in plant molecular biology is that from the cauliflower mosaic virus S35, which is very strongly expressed in plant cells, but this promoter does not provide the control over transcription required for successful expression. To express a foreign gene or to overexpress an endogenous gene it is essential to be able to control transcription. The best way to obtain this control is to use the plants own promoters, and the first step is to characterize these promoters.

Promoter sequences have been isolated and fused to reporter genes so that the expression pattern can be monitored easily. Many promoters that differ in their expression pattern with respect to tissues, environmental conditions, or developmental stages have been characterized following this approach.

Another problem in protein expression is how to obtain a subcellular targetting. The plant cells (like all eukaryotic cells) consist of several compartments (e.g., nucleus, chloroplasts, mitochondria). The proteins that are expressed in compartments other than the cytosol usually have signal sequences that direct them to their subcellular destination. To direct a protein to the desired compartment, a fusion is effected between the signal sequence and the mature protein. Many such signal sequences have been identified and are available to direct proteins to practically any compartment.

VII. ANTISENSE TECHNOLOGY

Genes encoding enzymes involved in starch biosynthesis or other relevant pathways (e.g., synthesis of sucrose) can be used for the overexpression of enzyme activity, as described in this chapter for ADPGlc PPase. Another approach is the use of antisense (complementary) DNA or RNA to decrease gene expression, a good way to assess the role of an enzyme and whether it limits the rate of the overall pathway.

A chimeric gene encoding antisense RNA for ADPGlc PPase, reducing the expression of the enzyme (Müller-Röber *et al.*, 1992) to between 2 and 5% of the wild type, reduced starch content of potato tubers by the same percentage; the number of tubers increased, but their weight decreased.

In wild type potatoes the amylose content varies from 18 to 23%, but potato plants with altered starch composition and content have been obtained by a number of different approaches. By using the antisense technology for genes encoding granule-bound starch synthase (GBSS) and branching enzyme, potato starches with different ratios of amylose and amylopectin were obtained (Visser *et al.*, 1991). Using antisense technology to decrease the expression of Waxy protein, it was seen that the amount of amylose deposited in the starch granule was related to the activity of the GBSS protein. The little amylose present in starch granules from such tubers was shown to be located at the hilum of the granule in a core of varying size that is surrounded by amylose-free starch (Visser *et al.*, 1991).

To test the hypothesis that phosphate supply from the can limit the rate of photosynthesis (Sivak and Walker, 1986), antisense experiments were performed by Schultz *et al.* (1993). A cDNA for the potato triose phosphate translocator was identified and a fragment of this cDNA in reverse orientation was expressed in trangenic potato plants under the control of the constitutive cauliflower mosaic virus 35S promoter (Rismeier *et al.*, 1993). This experiment confirmed that P_i supply can limit photosynthesis since a reduction of just 30% of P_i transport activity resulted in a decrease of 50% in the maximum rate of photosynthesis. As predicted (Sivak and Walker, 1986), this limitation if P_i supply also resulted in an accumulation of transitory starch in the leaves.

VIII. OTHER USES OF GENE TECHNOLOGY

Foreign enzymes—that is, with no plant equivalent, have been introduced into plants; one of the uses of this approach was to address the nature of sucrose transport into the phloem. An invertase derived from the yeast enzyme was targetted to the cell wall of tobacco, potato, tomato, and *A. thaliana* (Sonnewald *et al.*, 1994). The introduction of invertase decreased yield, presumably through the inhibition of sucrose transport.

The inorganic pyrophosphatase of *E. coli* was expressed in the cytosol of transgenic tobacco and potato plants, using the constitutive promoter of 35S cauliflower mosaic virus and the poly-A site of the octopine synthase gene terminator. Pyrophosphatase activity increased twofold in the transgenic plants relative to the controls, and the concentrations of pyrophosphate and pyruvate (indicating flow of photosynthates towards glycolysis) decreased. Sucrose content increased more than tenfold in the source leaves of the transgenic tobacco plants.

It should be noted that it is not enough to increase the amount of enzyme protein in a plant tissue to ensure higher activity of the enzyme. This is because individual enzymes may be regulated by several metabolites and the concentration of these cannot be controlled easily. Also, plant metabolism is integrated in ways that may escape control by the plant technologist, and altering one pathway may somehow affect alternative pathways in unpredictable ways. For example, Zrenner *et al.* (1993) managed to reduce the amount of UDPGlc PPase in potato tuber to just 4% of the wild-type level without any visible effect on carbohydrate metabolism.

Genetic modification of many dicotyledoneous crops such as potato using gene transfer via *A. tumefaciens* is, at present, an efficient and reliable technique; the application of the antisense route to limit or neutralize the action of undesirable genes has also been applied successfully to potatoes. Conversely, for monocotyledoneous crops such as maize, wheat, and rice, transformation can be accomplished by the much less efficient particle gun technique. This technique has been improved, resulting in rapid progress in the development of maize and later wheat with improved agronomic properties and/or altered starch composition.

Modification of wheat is also hampered for commercial reasons: unlike maize, wheat seeds have no male sterility, allowing the farmer to obtain new seeds from his wheat harvest. Introduction of male sterility into wheat by genetic engineering would make wheat breeding as attractive economically as maize breeding. In wheat the large variation in starch granule size is a negative factor for optimal use of wheat starch, a characteristic that is unlikely to be modified by conventional breeding and is awaiting a more original approach.

The information provided in this chapter is limited not only by matters of space and relevance, but also by an additional reason. A number of biotechnological companies and institutions are working in the development of new cultivars with altered starch compositions. Research done by, or on behalf of, commercial enterprises generally is not published in refered journals and is kept secret by the companies until applying for a patent. This policy of secrecy clearly slows down the dissemination of scientific information and deprives scientists of the very helpful peer review (affecting negatively the quality of the research). This tendency, unfortunately, is likely to become more dominant. Limited information is sometimes offered in scientific conferences, but hard data are often missing.

Most of our knowledge on storage starch and its biosynthetic enzymes comes from crop plants that have been genetically manipulated to increase starch content for thousands of years through plant breeding. It would not be surprising, then, if some of the peculiarities of the genetics of ADPGlc PPase (number of gene families; even variations in subcellular localization) were "artifacts" introduced by the selection towards high yield and high starch selection pressure by humans.

IX. TRANSFORMATION OF PLANTS WITH AN *Escherichia coli* ALLOSTERIC MUTANT *glg* C GENE INCREASES STARCH CONTENT

As discussed in the chapter, "The Biosynthetic Reactions of Starch Synthesis," there is a preponderance of evidence indicating that the rate-limiting and regulatory enzyme of starch synthesis in algae or bacterial glycogen synthesis is the ADPGlc PPase. With respect to higher plants, control analysis experiments have shown that ADPGlc PPase is important in the regulation of leaf starch synthesis (see the chapter, "Starch Accumulation in Photosynthesis Cells). Also, reduced ADPGlc PPase activity in mutants led to a reduction in the rate of starch synthesis in potato tubers (Müller-Röber *et al.,* 1992). Therefore, it was of interest to see if the starch content in a plant could be augmented by increased expression of activity of one of the enzymes involved in starch biosynthesis. Overexpression of a plant ADPGlc PPase activity, however, would require the expression of two distinct genes to reconstitute its ADPGlc PPase activity. Moreover, it is possible that the plant would compensate for the overexpression by altering the ratio of the effector metabolites, 3PGA and P_i, so that starch synthesis would not increase. Thus, a different strategy was chosen: an *E. coli* ADPGlc PPase, *glg* C gene of allosteric mutant 618, referred to as *Glg* C16 (Leung *et al.*, 1986), which encodes for an enzyme independent of the presence of activator for activity was used for the transformation. Expression of the bacterial mutant gene would have two advantages. Only one gene has to be expressed for ADPGlc PPase activity, and the mutant enzyme would be less sensitive to inhibition by its allosteric inhibitor, 5′AMP, insensitive to the inhibitor of the plant enzyme, P_i and independent of the activator for good activity (Leung *et al.*, 1986) (Table I). A collaboration with the Monsanto group was initiated to transfect plant systems with *Glg* C16 to see if the starch content of plants could be increased (Stark *et al.*, 1992). Because starch synthesis occurs in the plastid, a nucleotide sequence encoding the transit peptide of the *A. thaliana* ribulose 1,5-bisphosphate carboxylase chloroplast transit peptide was fused to the translation initiation site of the *glg* C16 gene (Fig. 1). A promoter was also needed; the chimeric gene was cloned behind a cauliflower mosaic virus (CaMV)-enhanced 35S promoter, or a tuber-specific patatin promoter, or, in the case of tomato plants, the *Arabidopsis* plant promoter from the *rbc*S gene was used (Stark *et al.*, 1992; Fig. 1). A polyadenylation signal from the nopaline synthase gene (Nos) was fused on at the 3′ end of the chimeric gene. The chimeric gene-containing promoter was placed in a cloning vector with a 35S-neomycin phosphotransferase gene as a selectable marker (Stark *et al.*, 1992) and was used for the transformation of tobacco calli, tomato cotyledons, and potato plants.

TABLE I

ACTIVATION AND INHIBITION OF THE ADPGLC PPASE ACTIVITY PRESENT IN TOBACCO PROTOPLASTS TRANSFORMED WITH *glg* C16[a]

Source of protoplast extract additions to assay	ADPGlc formed (nmol)
Nontransformed cells + 10 m*M* inorganic phosphate	0.0
Transformed + 2.5 m*M* fructose 1,6-bis-P	20.2
Transformed + 2.5 m*M* fructose 1,6-bis-P + 10 m*M* inorganic phosphate	18.0
Transformed + 20 m*M* + 3-P-glycerate	18.4
Transformed + 10 m*M* inorganic phosphate	6.4

[a] Data from Stark *et al.* (1992).

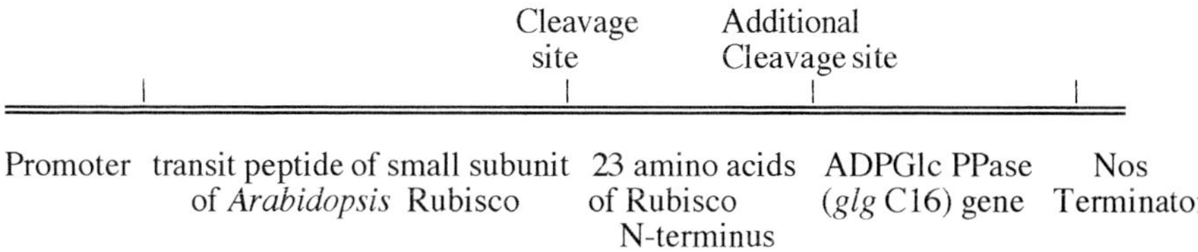

FIG. 1. Synthetic promoter–plastid transit peptide-*glg* C16 ADPGlc PPase gene. The chimeric gene is composed of the *Arabidopsis thaliana* chloroplast transit peptide portion of the ribulose bis-P carboxylase gene modified to have an extra cleavage site to eliminate the 23 amino acids of the N-terminal of the small subunit (Stark *et al.*, 1992) to prevent its possible interference with the catalytic or regulatory activity of the *glg* C16 gene product. The Nos terminator is the nopaline synthase 3′ poly A signal. The promoter can be either a constitutive promoter or a tissue-specific promoter.

In tobacco calli where the *glg*C gene product activity was detected, starch content was 1.7 to 8.7 times higher than in the controls lacking the *glg*C gene product (Stark *et al.*, 1992). The CaMV–chimeric gene was electroporated into tobacco protoplasts, and extracts of the transformed protoplasts gave rise to ADPGlc synthesis resistant to P_i inhibition and activated by fructose 1,6-bis-P (Table I). The synthesis of ADPGlc in the control protoplast extract was totally inhibited by P_i as expected since the tobacco and almost all plant ADPGlc PPases are most sensitive to inhibition by P_i. When the transgenic tobacco was examined by light microscopy and was compared with control calli, it showed a large increase in the number of starch granules (Stark *et al.*, 1992). Similarly, when tomato was transformed, with a construct transit peptide-*Glg* C16 gene, shoots excised from calli stained black with iodine reagent, whereas the controls were essentially negative.

Similar results have been obtained for Russet-Burbank potato tubers in which the chimeric gene, with its transit peptide under the control of a tuber-specific patatin promoter, increased starch in the tuber 25 to 60% over controls not containing the bacterial enzyme (Stark *et al.*, 1992; Table II). If the bacterial ADPGlc PPase *Glg* C16 gene was expressed in the tuber lacking the transit peptide gene portion, no increase in starch content was noted (Table II). Probably, ADPGlc PPase was expressed, but was not present in the amyloplast and, for this reason, was not able to supply ADPGlc to the starch synthases that are localized in the amyloplast.

A positive relationship between the expression levels of the ADPGlc PPase of *Glg* C16, as measured by immunoblotting of the potato extracts, and the increase in starch content was demonstrated, particularly in tubers at lower ranges of starch content. Lower levels of the expressed ADPGlc PPase resulted in increases of 21 to 63% in starch, intermediate levels of the expressed ADPGlc PPase gave increases of 33 to 118% in starch, and

TABLE II

STARCH CONTENT IN POTATO TUBERS TRANSFORMED WITH THE *glg* C16 AND *glg* C GENES[a]

	Plasmid used for transformation	Average starch content (% fresh weight)
A	Control; untransformed	12.3 ± 1.15
	Chloroplast transit peptide-*glg* C16	16.0 ± 2.00
	glg C16, no transit peptide	12.4 ± 0.24
B	Control; untransformed	13.2 ± 0.12
	Chloroplast transit peptide-*glg* C	13.1 ± 0.07

[a] Data from Stark *et al.* (1992).

the high expressed levels of the transit peptide-*Glg* C16 resulted in increases of 33 to 167%. It is of interest that when the wild-type *E. coli* ADPGlc PPase gene was expressed in the tuber, no increase in starch was noted (Table II). Thus, an important factor in increasing starch synthesis is to transform the tuber with an ADPGlc PPase with allosteric properties optimized to permit higher rates of ADPGlc synthesis under physiologic conditions.

X. ARE OTHER STARCH BIOSYNTHETIC ENZYMES RATE LIMITING?

Smith (1988) showed that in mutant *rr* pea leaves, in high light intensity, there was a 40% decrease in the rate of starch synthesis. A control coefficient analysis reported later (Smith *et al.*, 1990) showed that in low light intensity, there was essentially no effect on the rate of starch synthesis, whereas in high light intensity, the flux control coefficient value was 0.13, which is a small value (meaning very little control) and is only one-fifth the value seen for ADPGlc PPase (Neuhaus and Stitt, 1990). Thus an 86% reduction of branching enzyme activity had a small effect on regulation of starch synthesis.

It has been suggested that when plants are subjected to high temperature, starch synthase activity may be rate limiting. At temperatures higher than 30°C, both maize (Singletary *et al.*, 1994) and wheat endosperm (Hawker and Jenner, 1993; Keeling *et al.*, 1993, 1994; Jenner, 1994) had a reduction of starch deposition as compared with lower temperatures. In wheat, the starch biosynthetic enzyme affected was soluble starch synthase (SSS).

Using flux control coefficient analysis, Keeling *et al.* (1993) showed a control coefficient close to 1 between the rate of starch synthesis and the level of starch synthase activity in wheat endosperm extracts. It was also shown that in vitro, the endosperm starch synthase activity was sensitive to heat treatment in the range of 30 to 40°C if the treatment was for longer than 15 minutes. A similar study with maize endosperm showed a reduction of starch synthetic rate and a decrease in starch synthase activity in the heat-stressed maize endosperm (Singletary *et al.*, 1994). It was also noted, however, that in the heat-stressed maize, the endosperm ADPGlc PPase activity was also reduced to an even greater extent than the SSS (Singletary *et al.*, 1994).

Thus, in wheat and maize, under some environmental conditions, there might be a correlation between reduction of starch synthase activity and decreased starch synthesis. However, as the data obtained with maize suggest (Singletary *et al.*, 1994), other unknown factors, beside starch synthase activity, may be the primary reason for the reduction of starch synthesis in the heat-stressed plants. In the case of maize endosperm, another enzyme involved in starch synthesis. ADPGlc PPase, is also affected in the heat-stressed plant. It is also possible that other critical steps leading to starch biosynthesis are affected in both plants, such as carbon flow from source to sink tissues and invertase activity. Those processes were not studied in the heat-stressed plants. Thus, we believe that the published evidence does not warrant the designation of starch synthase as a major control point. Flux control coefficients for an enzyme within a process can only be determined if the activity of only that enzyme is affected. In the case of heat-stressed plants, it has not yet been shown that only the starch synthase activity is affected. A crucial test is whether the starch synthetic rate can be increased by overexpressing soluble starch synthase activity in the amyloplast. As shown in the preceding, starch accumulation can be increased by expressing a bacterial ADPGlc PPase allosteric mutant in plants (Stark *et al.*, 1992).

XI. OTHER PHYSIOLOGIC EFFECTS OF MANIPULATION OF STARCH SYNTHESIS

Starch phosphorylase and amylolytic enzymes are responsible for starch degradation during cold storage of potato tubers (see the chapter, "Starch Degradation"), and result in the formation of glucose-1-P and glucose from starch. Glucose-1-P may also be formed from the products of degradation of sucrose via invertase or sucrose synthase. Sugar accumulation, or "sweetening," decreases the quality of the tubers and makes them unsuitable for frying. Also, accumulation of sugars is eventually followed by the end of dormancy, which is signaled by the onset of respiration and sprouting. Barry *et al.* (1994) found that by overexpressing ADPG PPase during cold storage they could delay cold sweetening and sprouting. They proposed that ADPG PPase would act as an active sink for the glucose-1-P and glucose (which can be converted into glucose-1-P) products of starch degradation and the ADPGlc formed is then converted back into starch. Overexpression of ADPG PPase during cold storage is achieved by a variation of the method used by Stark *et al.* (1992) and by using an *A. thaliana* cold-inducible promoter. The potato tubers obtained in this manner had better frying properties than the control after cold storage.

Giroux *et al.* (1996) described the effect of a single gene mutation in the *shrunken* 2 locus of maize (coding for the large subunit of the ADPGlc PPase), which involved the addition of 2 amino acids—tyrosine and serine. The mutation decreased the sensitivity of the enzyme to inhibition by phosphate and was introduced by using an in vivo, site-specific mutagenesis system that involved the use of the transposable element *ds* (dissociation). The mutated gene, named Rev6, increased seed weight by 11 to 18% without changing the proportion of the seed weight taken by starch. The authors proposed that increased ADPGlc PPase activity would affect the overall sink strength of the seeds, as it increased not just starch content but also other constituents of the seed.

XII. CONCLUSIONS

It is conceivable that methodology such as that used to increase starch quantity could be used to influence starch quality by manipulation of starch synthase and branching isoforms. These "new starches" may have greater usefulness in food and industrial processes. The production of modified "specialty" starches via molecular biology techniques is promising, and perhaps more beneficial and more economical than the chemical modification of starch for industrial purposes.

Because there has been an increased demand for starch for both specialized industrial and food uses since the mid-1980s (Katz, 1991), it appears that the study of basic questions on the structure–function relationships of the allosteric regulation of an enzyme involved in sugar nucleotide synthesis now may have a great impact on both agriculture and industry. It is of interest that this research on the routes and mechanism of regulation of bacterial glycogen and starch synthesis at the molecular level, which began in the mid-1960s, has led to opportunities for improving the quality of the uses for starch in industrial and food processes. This was never the original purpose of the studies, but is an example of how basic science, which tries to answer basic questions, may lead to methods where nature can be manipulated for beneficial purposes.

In addition to altering quantity, starch quality could be changed via expression of the isoforms of starch synthase and branching enzymes in plants. These "designer" starches would be used in the food and other industries. One possible approach to modifying starch structure in a crop would be the replacement of the plant branching isozymes with other different properties. The foreign enzyme could be a chimeric construct.

Although construction of chimeric enzymes is done to elucidate the domains that determine the different properties of the isoforms (see the chapter, "Branching Enzymes"), a secondary benefit is that novel enzymes may be capable of branching starch differently than the wild-type enzymes, resulting in the production of starch with novel properties.

FURTHER READINGS

These sources provide additional in-depth coverage of this topic. For complete reference, please see the Reference section at the end of the book.

Lea, P. J., and Leegood, R. C. (1993)
Marcus, A. (1989)
Walden, R., Koncz, C., and Schell, J. (1990)

STARCH ACCUMULATION IN PHOTOSYNTHETIC CELLS

I. INTRODUCTION

The classic equation for photosynthesis is often written as

$$6CO_2 + 6H_2O \rightarrow C_6H_{12}O_6 + 6O_2 \tag{1}$$

The most likely end product, designated in the equation by $C_6H_{12}O_6$, is not actually hexose. Originally, starch seemed the most likely end product, and the historic observations of Sachs, Pfeffer, and Godlewski established an intimate relationship between photosynthesis and the process of starch accumulation in green leaves (see Rabinowitch, 1945). However, even in the time of Sachs, it was known that some plants do not accumulate starch within their leaves under any circumstances and, in addition, an almost insoluble polysaccharide could clearly not be moved about the plant. Eventually, it became increasingly accepted that sucrose was the real end product of photosynthesis (Rabinowitch, 1956), whereas starch was relegated to the role of a temporary storage compound. Even when it became evident that sugar phosphates played a central role in photosynthetic carbon metabolism (Benson *et al.*, 1950, 1952; Benson and Calvin, 1950; Bassham and Calvin, 1957), sugar phosphates were regarded as intermediates. Conversely, the percentage of radioactive carbon in sucrose extrapolated to zero at zero time, and the percentage increased thereafter in a way that might have been predicted for an end product awaiting movement to other parts of the plant.

The rate of photosynthesis does not depend on the amount of a single component (e.g., the activity of a particular enzyme). There is a wide range of possible regulatory factors, proven to exist in vitro, but the importance of which in vivo has still to be determined. In particular, there is a multitude of factors affecting the activity of the enzymes involved, with pH, ions, coenzymes, and metabolite effectors modulating the activity of every enzyme studied thus far. Compartmentation is the other key factor. The role of metabolite transport in the cell, particularly between chloroplast and cytosol, but also to and from mitochondria, vacuole, and other organelles, is now considered to be fundamental to the regulation of photosynthesis. In this chapter, we look at the factors considered to be of major importance in the determination of the nature of the products of photosynthesis, and we look at the partition of carbon and energy between sucrose and starch.

II. THE REDUCTIVE PENTOSE PHOSPHATE PATHWAY

The reductive pentose phosphate pathway (RPPP), also called the Benson-Calvin cycle, is the only pathway in plants that can catalyze the net fixation of CO_2. The entire cycle can be divided into three phases (Fig. 1):

1. Carboxylation of ribulose-1,5-bisphosphate by ribulose 1,5-bisphosphate carboxylase-oxygenase (Rubisco) with the formation of two molecules of glycerate-3-P.
2. Reduction of the two molecules of glycerate-3-P to triose phosphate at the expense of 2 ATP and 2 NADPH.
3. Regeneration of the primary acceptor, RuBP, from triose phosphate in the sugar phosphate shuffle.

Autocatalysis is a crucial property of the RPPP, and it refers to the fact that the product, triose phosphate, can be recycled, generating more substrate for carboxylation. If the cycle turns over five times, the amount of the primary acceptor, RuBP, doubles. During steady-state photosynthesis (after the induction is over), one-sixth of the triose phosphate generated from CO_2 is available for product synthesis. There is, however, a relatively major drain on fixed carbon because photorespiration results in the net loss of carbon and energy. Photorespiration refers to the fact that C_3 plants evolve CO_2 when illuminated in CO_2-free air (apparently, it is absent in C_4 plants). Glycollate is considered the primary substrate of this "light respiration"; Rubisco is a branch point between photorespiratory and photosynthetic metabolism. Oxygen, reacting with RuBP, leads to glycollate synthesis and photorespiration, whereas CO_2, reacting with RuBP, leads to photosynthesis. The glycollate formed by Rubisco when it oxygenates RuBP cannot be used directly in the RPPP, although it is salvaged to some extent by the photorespiratory pathway.

Some of the triose phosphate available for product synthesis will be transported into the cytosol and converted into sucrose (Fig. 2). In both photosynthetic and nonphotosynthetic cells, sucrose is synthesized via sucrose phosphate synthase, which catalyses the reaction

$$\text{UDP-glucose} + \text{fructose-6-phosphate} \rightarrow \text{sucrose-6-phosphate} + \text{UDP}.$$

The sucrose phosphate formed is hydrolyzed by a specific phosphatase to

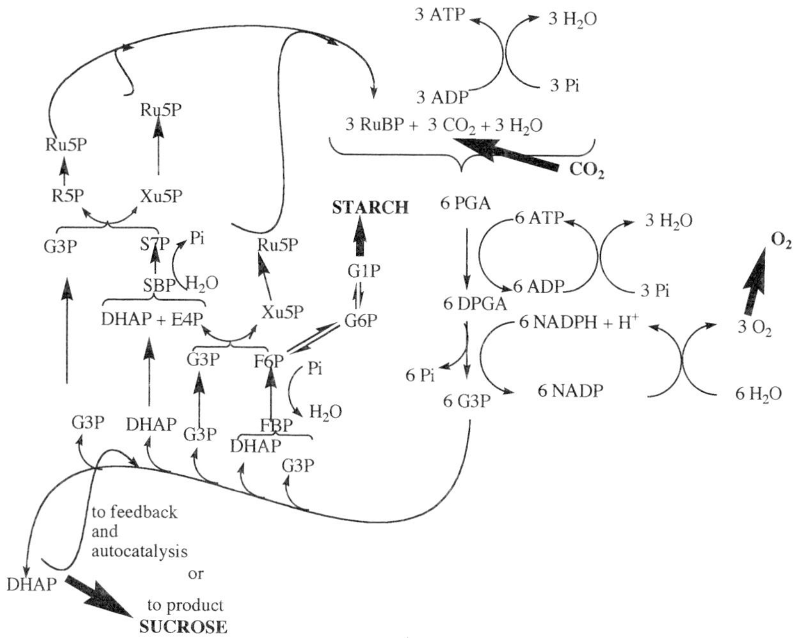

$$3\ CO_2 + 2\ H_2O + Pi \longrightarrow CH_2O\text{-}CO\text{-}CH_2OPO(OH)_2 + 3\ O_2$$

FIG. 1. The reactions that lead to the regeneration of RuBP and the formation of triose phosphate. On the right, three molecules of RuBP combine with three molecules of CO_2 and three molecules of water to give six molecules of PGA. These are phosphorylated at the expense of ATP, and the resulting DPGA is reduced by NADPH to G3P. The major part of this is converted to its isomer DHAP. Aldol condensation of these two triose phosphates give a molecule of FBP, which undergoes hydrolysis to F6P. This hexose phosphate is also the precursor of G6P and G1P, which, after further transformation, give rise to starch. The F6P also enters the first transketolase reaction donating a 2-carbon unit to G3P to form Xu5P and E4P. The process of condensation, phosphorylation, and 2-carbon transfer is repeated, yielding SBP, S7P, and two more molecules of pentose phosphate, respectively. All three molecules of pentose monophosphate are finally converted to Ru5P, which is phosphorylated to RuBP.

give free sucrose. The formation of UDPglucose is analogous to the formation of ADPglucose in starch synthesis.

It is worth noting that the two sucrose metabolizing enzymes, sucrose synthase and sucrose-phosphate synthase, were both discovered by Leloir and Cardini (1955) (see "Preface").

There is yet another possible route for the triose phosphate formed in the RPPP, and that is starch synthesis within the chloroplast. Stromal starch

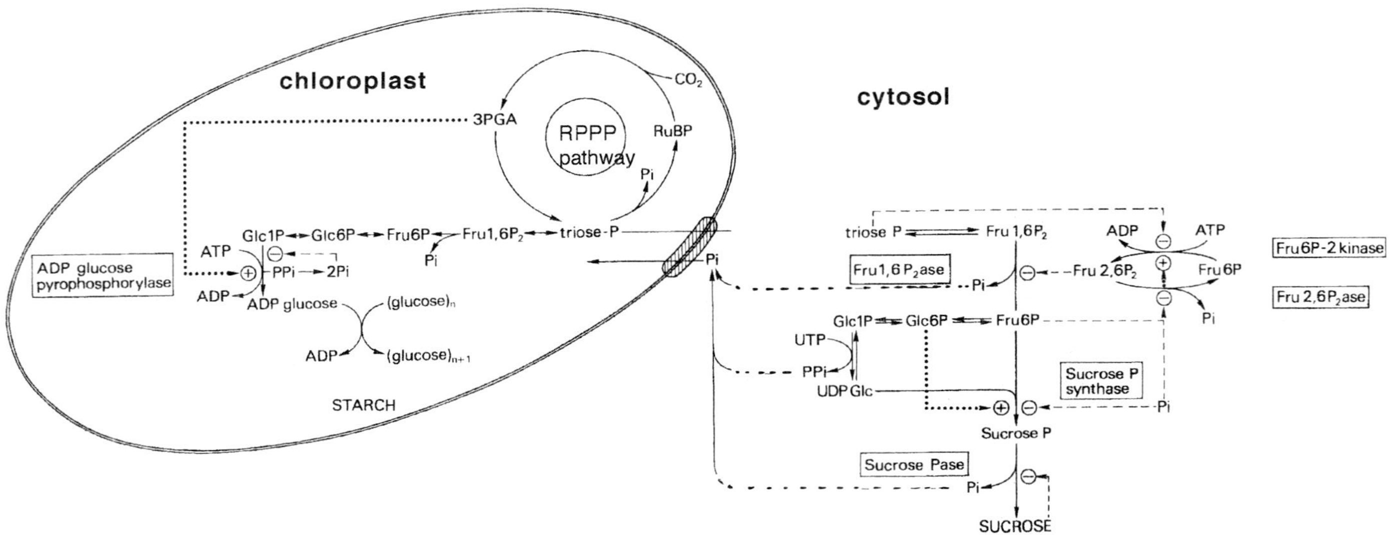

FIG. 2. The control of synthesis of sucrose and starch in photosynthetic cells, and the role of metabolite modulation, including that by fructose 2,6-bisphosphate.

is formed primarily from triose phosphate released from the RPPP. If triose phosphate is retained within the chloroplast, the initial reactions are the same as those involved in sucrose synthesis in the cytoplasm (i.e., a proportion of the triose phosphate undergoes aldol condensation to Fru1,6P_2, which is then hydrolyzed to Fru6P). This is then converted into its isomer by hexose phosphate isomerase. In the reaction catalyzed by phosphoglucomutase, Glc6P is converted into Glc1P. At equilibrium, a mixture of the enzymes mentioned previously would yield hexose phosphates in the proportions of approximately Fru6P (9) to Glc6P (17) to Glc1P (1). Although these reactions are considered freely reversible, the overall equilibrium may still be important in determining the distribution of carbon between starch and pentose monophosphate in the illuminated chloroplast. Both Fru6P and triose phosphate are substrates for the first transketolase reaction, and this, in turn, influences the amount of triose phosphate entering the second aldolase condensation and the second transketolase reaction. An active sink for Glc1P would therefore tend to deflect carbon toward starch. This sink could be provided by ADPGlc PPase, which catalyses the reaction

$$\text{glucose-1-P} + \text{ATP} \rightarrow \text{ADPglucose} + \text{PP}_i$$

when low external P_i decreases triose phosphate and PGA export.

Few studies on the localization of the starch biosynthetic enzymes were done before 1978, when it was found that ADPGlc PPase was located exclusively in the chloroplast fraction in both spinach (Mares *et al.,* 1978) and pea (Levi and Preiss, 1978). The first detailed study was done by Okita *et al.* (1979), in which spinach leaf chloroplasts were isolated either by differential centrifugation (Walker, 1971; see also later) or from protoplasts (Nishimura *et al.,* 1976). These plastid preparations contained essentially all of the activity of the starch biosynthetic enzymes, ADPGlc PPase, starch synthase, and branching enzyme. Subsequently, in guard cells of *Commelina communis,* Robinson and Preiss (1987) showed that the starch biosynthetic enzymes were present exclusively in the chloroplast fraction.

III. THE CHLOROPLAST AS A TRANSPORTING ORGANELLE

The chloroplast must operate its carbon cycle as an autocatalytic breeder reaction, but it must export elaborated carbon and chemical energy to its cellular environment. In order to export, it must produce more than it uses, but can only do this by returning newly synthesized intermediates to the cycle. In order to satisfy the needs of the cell, it must release newly made products to the cytoplasm. These competing processes can be accomplished efficiently because the choloroplast keeps a delicate balance among recycling, export, and internal storage.

Chloroplasts are enclosed by two membranes. The outer membrane is freely permeable to small molecules (up to about 10 kDa) due to the presence of a *porin* and the inner membrane is the osmotic barrier and the site where specific transport occurs. The specificity of envelope permeability is strikingly highlighted by the contrast between P_i and PP_i, the former being among the most rapidly translocated molecules and the latter among those to which the envelope is relatively impermeable. Carrier-mediated anion transport can be classified as:

1. Electroneutral, involving exchange of one anion with another of equal charge
2. Electroneutral proton compensated, in which the different charge is compensated by cotransport of a proton
3. Electrogenic, involving an exchange between anions of different charges (which requires energy as membrane potential or proton electrochemical gradient)

The study of transport by isolated chloroplasts requires the use of "good" organelles, and the criteria for this are photosynthetic rate and chloroplast intactness. If isolated chloroplasts are capable of rapid electron transport and photophosphorylation, but have lost the ability to assimilate CO_2 when illuminated in a suitable reaction mixture, they have then been damaged or have been irreversibly inhibited during isolation. Because there is often a clear correlation between envelope integrity and function, results obtained with relatively inactive chloroplasts are unlikely to reflect the behavior of chloroplasts in situ.

Techniques developed with the aim of separating intact chloroplasts from leaf tissues (Walker, 1971) yield preparations containing, on average, some 70 to 80% (or higher, depending on the species and quality of the material) of class A chloroplasts (Hall, 1972).

IV. CONTROL OF CARBOHYDRATE METABOLISM

During the day, the rates of starch and sucrose synthesis and the rate of photosynthetic carbon assimilation must be coordinated. There is a clear need to determine how much assimilated carbon can be diverted into sucrose and starch synthesis without decreasing too much the amount that returns to the RPPP. Conversely, when sucrose accumulates in the cytosol because the rate of export diminishes (and/or photosynthesis increases), starch begins to accumulate inside the chloroplast. During the night, the sucrose accumulated in the vacuole during the day and the starch accumulated in the chloroplast are remobilized to be used to support the metabolism of the leaf itself or to be exported as sucrose (Stitt *et al.*, 1987a). Stromal amylases and phosphorylases degrade starch, and Glc1P is then converted into triose phosphate, which can be exported from the chloroplast.

In this way, photosynthates are constantly available. The importance of these remobilization mechanisms is highlighted when they are disturbed. For example, mutants of the crucifer *Arabidopsis thaliana,* which are unable to synthesize starch but can still synthesize sucrose, grow at the same rate as the wild type under continuous light, but the growth rate is drastically diminished if placed in a day–night regime (Caspar *et al.*, 1986).

V. REGULATION OF THE ADPGlc PATHWAY IN THE CHLOROPLAST

As discussed elsewhere (see the chapter, "Synthesis of the Glucosyl Donor: ADPglucose Pyrophosphorylase"), for every leaf system studied, whether the leaf source is from a plant using the C_3 or C_4 pathway or Crassulacean metabolism, the major activator is still 3PGA and the inhibitor is P_i. There is much evidence obtained in vitro suggesting that ADPGlc synthesis is regulated by activation of the plant ADPGlc PPase by 3-phosphoglycerate (3PGA) and inhibition by P_i. In vivo and in situ experiments showed a correlation between the concentrations of 3PGA and starch, and inverse correlations between P_i and starch levels (see the chapter, "Regulation of the Starch Synthesis Pathway: Targets for Biotechnology").

The increasing availability of mutant and transgenic plants now facilitates the study of how plant metabolism is controlled. In particular, control analysis involves asking how much a flux changes for a given change in enzyme activity, such that the flux control coefficient.

$$C_{JE} = \frac{dJ/J}{dE/E}$$

In this equation, E is the original amount of enzyme, J is the original pathway flux, and dJ is the change that results from a relatively small change in the amount of the enzyme dE. For an enzyme in a simple, unbranched pathway, if $C_{JE} = 1$, then C_{JE} can vary between 0 (no control) to 1 (total control), that particular enzyme limits the rate of the overall pathway (see Kacser and Burns, 1973; Heinrich and Rapoport, 1974).

The availability of chloroplast mutants of phosphoglucose isomerase of *Clarkia xantiana* (Kruckeberg *et al.*, 1989; Neuhaus *et al.*, 1989), of phosphoglucomutase (Caspar *et al.*, 1986), and of ADPGlc PPase of *A. thaliana* (Lin *et al.*, 1988a,b; Neuhaus and Stitt, 1990), has allowed the analysis of the extent of control that these enzymes exert on chloroplast starch synthesis. Mutant plants with reduced activity of both cytosolic (64%, 36%, 18% of wild type) and chloroplastic (75%, 50% of wild type) phosphoglucoisomerase were used to determine the effect of these enzymes on fluxes toward starch and sucrose synthesis as well as on photosynthetic rate and control coefficients (Kacser and Burns, 1973; Kruckeberg *et al.*, 1989). The plastid phosphoglucoisomerase exerted little control over starch or sucrose synthesis in low light, but did exert control of starch synthesis in saturating light. Lowering the cytosolic enzyme activity had little effect on either starch or sucrose synthesis in saturating light, but increased starch synthetic rate and decreased sucrose synthesis in low light. Thus variation of the cytosolic phosphoglucoisomerase affected the partitioning of carbon between sucrose and starch. Further studies (Neuhaus *et al.*, 1989) confirmed that reduction of plastid phosphoglucoisomerase had little effect in low light, but reduced starch synthesis by 50% in saturating light with no corresponding increase in sucrose synthesis. Reduced levels of cytosolic enzyme (18% of wild type) lowered the sucrose synthetic rates and increased the rate of starch synthesis. Metabolite levels were also affected in these mutants. In the mutant containing only 18% of the wild type cytosolic phosphoglucoisomerase activity, both fructose-2,6-bisphosphate and 3PGA levels increased approximately 100%. Neuhaus *et al.* (1989) suggested that the lower rate of sucrose synthesis rate is due to the increased Fru-2,6-P_2 concentration, which causes increased inhibition of cytosolic fructose-1,6-bisphosphatase (for reviews on sucrose synthesis and its regulation, see ap Rees, 1987; Stitt *et al.*, 1987b), which is on the pathway toward sucrose synthesis (Fig. 2). Their data strongly support the view that increased starch synthesis in the mutants with reduced levels of phosphoglucoisomerase is due to activation of the ADPGlc PPase by the increased 3PGA concentration and 3PGA/P_i ratio.

These experiments have been extended to the null chloroplast phosphoglucomutase (Caspar *et al.*, 1986) and the low activity (7% of wild type) ADPGlc PPase mutants (Lin *et al.*, 1988a,b) of *A. thaliana.* Neuhaus and Stitt (1990) used the alleles to construct hybrid plants containing, respectively, 50% of wild-type phosphoglucomutase activity and 50% of wild-type ADPGlc PPase activity. The effects of these reduced activities on starch and sucrose fluxes and on CO_2 fixation in low-light and high-light intensities were measured. In low light, a 50% decrease in phosphoglucomutase activity had no significant effect on the fluxes mentioned previously. However, a 50% and 93% decrease of ADPGlc PPase activity resulted in a 23% and 74% decrease in flux of starch synthesis, with a concomitant increase of a 17% and 42% increase in sucrose synthetic rate. Thus, a decrease in the synthesis of ADPGlc not only affected starch synthesis but also affected the partitioning of photosynthetic carbon, causing more to be directed toward sucrose biosynthesis. In high light a 50% decrease in phosphoglucomutase activity resulted in a 20% decrease in starch synthesis with little effect on the sucrose synthesis rate. However, reduction of the the ADPGlc synthesizing activity by 50% and 93% resulted in a 39% and 90% decrease in starch synthesis flux. The flux of photosynthetic carbon under these conditions was not redirected toward sucrose synthesis but rather the photosynthetic rate was inhibited approximately 46%. The flux control coefficients (Burns *et al.,* 1985) for the enzymes for starch synthesis were calculated to determine the distribution of control and were compared with previous results obtained with the *C. xantiana* phosphoglucoisomerase.

A kinetic model was developed by Petersson and Ryde-Petersson (1989) that was consistent with the metabolite concentrations and mass action ratios measured in vivo and with enzyme properties and equilibria. These authors reached the conclusion that 3PGA and P_i play important roles in regulating starch synthesis with significant contributions made by ATP, glucose-1-P, and fructose-6-P. Since these metabolites are either substrates or effectors of the ADPGlc PPase, the analysis is consistent with the view that 3PGA is a positive effector and P_i is a negative effector of ADPGlc synthesis and, therefore, that the 3PGA/P_i ratio regulates starch synthesis via regulation of ADPGlc PPase.

In summary, analysis of the starch biosynthetic system in a number of plants or using data obtained in vivo from different plants and applying the control analysis method of Kacser and Burns (1973; see also Kacser, 1987) show that the major site of regulation of starch synthesis is at ADPGlc PPase and that 3PGA and P_i are important regulatory metabolites of that enzyme. A decisive proof that this regulatory mechanism is functional in vivo would be the isolation of a plant containing an ADPGlc PPase with altered allosteric properties that would correlate with its starch content. So far, such mutations have not been found in higher plants but have been reported for *E. coli* and *Salmonella typhimurium* (reviewed in Preiss and Romeo, 1989), and for *Chlamydomonas reinhardtii.* Ball and his collaborators at Lille have obtained mutants of this unicellular green algae that have an ADPGlc PPase with a low sensitivity to PGA activation; these mutants display a low starch content. The genetic manipulation of either the structural or regulatory genes of the starch biosynthetic enzymes may provide means for alteration of the starch levels in a plant, and a significant advance is described in the chapter, "Regulation of the Starch Synthesis Pathway: Targets for Biotechnology."

VI. STARCH SYNTHESIS IN YOUNG LEAVES

In young developing leaves that still behave as sinks (rather than sources), sucrose is first hydrolyzed in the cytosol by the action of the invertase or by the sucrose synthase followed by UDPGlc pyrophosphorylase. The hexose sugars formed are then metabolized via glycolysis into C_3 intermediates that are then transported into the chloroplast via the P_i translocator, where they can be used or stored as starch.

VII. SYNTHESIS OF STARCH AND SUCROSE IN C_4 PLANTS

The C_4 cycle can be viewed as an ATP-dependent CO_2 pump that delivers CO_2 from the mesophyll cells to the bundle-sheath cells, thereby suppressing photorespiration (Hatch and Osmond, 1976). The development of the C_4 syndrome has resulted in considerable modifications of inter- and intracellular transport processes. Perhaps the most striking development with regard to the formation of assimilates is that sucrose and starch formation are not only compartmented within cells, but in C_4 plants also may be largely compartmented between mesophyll and bundle-sheath cells. This has been achieved together with a profound alteration of the Benson-Calvin cycle function, in that 3PGA reduction is shared between the bundle-sheath and mesophyll chloroplasts in all the C_4 subtypes. Moreover, since C_4 plants are polyphyletic in origin, several different metabolic and structural answers have arisen in response to the same problem of how to concentrate CO_2. C_4 plants have three distinct mechanisms based on decarboxylation by NADP+-malic enzyme, by NAD+-malic enzyme, or by phosphoenolpyruvate (PEP) carboxykinase in the bundle-sheath (Hatch and Osmond, 1976).

Downton and Hawker (1973), showed that starch, starch synthase, and ADPGlc PPase were much higher in bundle-sheath cells than in mesophyll cells on a protein basis or on a chlorophyll basis. The mesophyll cell is able to synthesize starch on exposure of the leaf to continuous light for approximately 2.5 days. Under these conditions, starch synthase levels in the mesophyll cell increased. Thus the mesophyll cell is capable of starch synthesis under certain conditions. Later reports on other C_4 plants (e.g., nutsledge leaves, Chen *et al.,* 1974; *Digitaria pentzii,* Mbaku *et al.,* 1978) also indicate that both tissues are capable of starch synthesis.

Although starch synthase is present in both tissues, in *Digitaria pentzii* activity of the starch synthase is 10 times higher (calculated on a chlorophyll basis) in the bundle sheath. The studies on the maize leaf have been confirmed by more recent research, which is also more comprehensive since it also includes the measurement of branching enzyme activity (Preiss *et al.*, 1985; Echeverria and Boyer, 1986; Spilatro and Preiss, 1987). Their results are in agreement with the earlier conclusion that starch and the starch biosynthetic enzymes are primarily in the bundle sheath. Moreover, the exclusive localization of the starch biosynthetic enzymes in the bundle sheath chloroplast has been reported (Echeverria and Boyer, 1986). Thus, the discovery that the leaf starch biosynthetic is located solely in the chloroplast is in keeping with the finding that starch in higher plants is found exclusively in the chloroplast.

Although in a species such as maize the synthesis of sucrose appears to occur largely in the mesophyll cells: whereas the synthesis of starch occurs largely in the bundle-sheath cells, it is clear that there is a good deal of flexibility both within maize and between C_4 plants in general. For example, although the mesophyll tissue of maize grown under normal conditions contains no detectable starch, growth of plants in continuous light induces starch formation in the mesophyll (Downton and Hawker, 1973). However, *Digitaria* spp. (which, like maize, are also NADP+-malic enzyme–type C_4 plants) synthesize both sucrose and starch in the mesophyll compartment (Mbaku *et al.*, 1978; Hallberg and Larsson, 1983). A number of pieces of evidence support the contention that sucrose synthesis is confined to the mesophyll cells in leaves of at least some C_4 plants. Bucke and Oliver (1975) and Furbank *et al.* (1985) found that the majority of the sucrose–phosphate synthesis is located in the mesophyll of maize, *Pennisetum purpureum,* and *Muhlenbergia montana.* Other studies have shown the cytosolic fructose bisphosphatase to be confined largely to the mesophyll (Furbank *et al.*, 1985) as well as Fru6P, 2-kinase, fructose-2,6 bisphosphatase (Soll *et al.*, 1983), and fructose-2,6-bisphosphate itself (Stitt and Heldt, 1985). However, Ohsugi and Huber (1987) have shown that sucrose–phosphate synthase activity is present in both mesophyll and bundle-sheath cells in all C_4 subtypes, including maize. In addition, the response of the enzyme to light was different in the two compartments, with the bundle-sheath enzyme requiring higher irradiance for activation. Ohsugi and Huber (1987) suggest that sucrose–phosphate synthase may function in both mesophyll cells and bundle-sheath cells for sucrose synthesis in the light, particularly at high light intensity, whereas in the dark the major function of bundle-sheath cell sucrose–phosphate synthetase may be in sucrose formation following starch degradation, which is a function that has been largely overlooked. Perhaps the safest conclusion is that starch and sucrose synthesis may predominate in one or the other compartment in maize, but that this compartmentation may readily be overridden when environmental conditions (e.g., high light intensity) require it.

VIII. THE REGULATION OF STARCH SYNTHESIS IN C_4 PLANTS

A number of studies have shown that the activities of starch synthase, branching enzyme, and ADPGlc PPase are higher in the bundle-sheath cells than in the mesophyll cells. However, the enzymes of starch degradation, starch phosphorylase, and amylase are more evenly distributed and are slightly higher in the mesophyll (Huber *et al.,* 1969; Downton and Hawker, 1973; Echeverria and Boyer, 1986; Spilatro and Preiss, 1987). ADPGlc PPase from spinach is activated by 3PGA and inhibited by P_i, with ratios of 3PGA/P_i for half-maximal activation typically being less than 1.5 (Ghosh and Preiss, 1966). ADPGlc PPase from maize leaves requires much higher ratios of 3PGA to P_i for half-maximal activation, with the enzyme from the mesophyll cells requiring a higher ratio (9 to 16) than the enzyme from the bundle-sheath cells (7 to 10) (Spilatro and Preiss, 1987). Measurements show that 3PGA may be as high as 15 to 16 mol $\cdot$ m^{-3} in the bundle sheath and 5 to 7 mol $\cdot$ m^{-3} in the mesophyll (Leegood, 1985; Stitt and Heldt, 1985) due to the requirement for metabolite gradients during photosynthesis. Thus 3PGA/P_i ratios in the mesophyll are likely to be considerably lower than in the bundle sheath. This factor, and the relatively low activities of the enzymes of starch synthesis in the mesophyll, would appear to limit synthesis of starch in the mesophyll relative to the bundle sheath, but this factor is a relationship that could be modified readily with fluctuations in physiologic and developmental conditions, and could therefore account for variations in the capacity of the mesophyll to make starch.

IX. STARCH IN CAM PLANTS

Metabolism in CAM plants involves the transfer of large amounts of carbon between two storage pools: malic acid and storage carbohydrates (Fig. 3). Although malate is invariably stored in the vacuole, the carbohydrate store may be either chloroplastic (e.g., starch-storers such as *Bryophyllum tubiflorum, Kalanchoe diagremontiana*) or extrachloroplastic [as in *Ananas comosus* (pineapple) and *Aloe arborescens*]. This transfer of carbon involves glycolytic carbohydrate breakdown in the dark and gluconeogenic carbohydrate synthesis during the light. In starch-formers, large amounts of carbon must therefore enter the chloroplast during

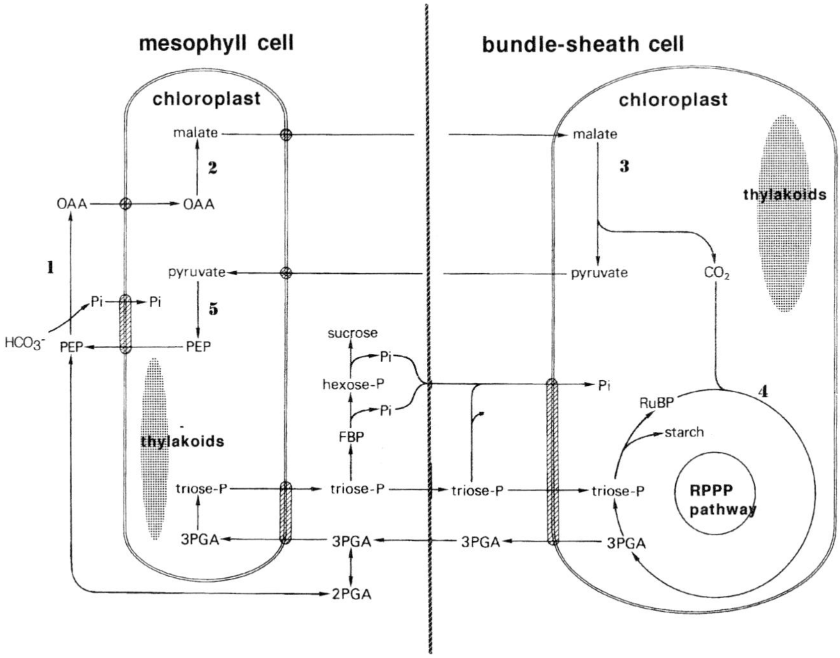

FIG. 3. Operation of the C_4 pathway in a NADP+-malic enzyme-type plant such as maize. The C_4 pathways, essentially CO_2 concentrating mechanisms, are classified according to the enzyme that decarboxylates the C_4 acid in the bundle-sheath chloroplast, decreasing the oxygenation reaction of Rubisco. In maize, the decarboxylating enzyme is a malic enzyme (3) that uses oxidized NADP (NADP+) as a cofactor. Hatched areas indicate carrier-mediated transporters of metabolites across the chloroplast envelope. Between cytosols of the two cell types, metabolites and phosphate move along diffusion-driven concentration gradients. For most reactions, cofactors and transaminations have been omitted for clarity. 1, phosphoenolpyruvate (PEP) carboxylase; 2, malate dehydrogenase; 3, malic enzyme; 4, Ribulose bisphosphate carboxylase/oxigenase (Rubisco); 5, pyruvate phosphate dikinase. In C_4 plants, the compartmentation and control of sucrose and starch synthesis is greatly modified with respect to the scheme shown in Fig. 2 in this chapter; for example, sucrose is synthesized mainly in the cytosol of mesophyll cells, and starch in the chloroplasts of bundle sheath cells.

deacidification. Fahrendorf *et al.* (1987) have proposed, as a working hypothesis, that in CAM plants (which are primarily starch-formers): 1. Malic enzyme is the principal decarboxylase, whereas pyruvate P_i dikinase is present in amounts sufficient to convert the pyruvate formed back into PEP; and 2. the capacity for conversion of FruBP into Fru6P in the cytosol is low and levels of Fru2, $6P_2$ are low during deacidification. It is suggested that this apparent paradox may be due to a very low affinity of the FruBPase for FruBP, as occurs in the CAM but not in the C_3 form of *Mesembryanthemum crystallinum* (Keiller *et al.,* 1987) nor in C_4 plants such as maize (Stitt and Heldt, 1985).

FURTHER READINGS

These sources provide additional in-depth coverage of this topic. For complete reference, please see the Reference section at the end of the book.

Edwards, G., and Walker, D. A. (1983)
Leegood, R. C. (1996)
Pontis, H. G., Salerno, G. L., and Echeverria, E. J., eds. (1995)
Quick, W. P., and Stitt, M. (1996)

STARCH DEGRADATION

I. PLANT AMYLASES AND PHOSPHORYLASES

Many starch-degrading enzymes have been isolated from fungi, yeasts, and bacteria, and their mode of action varies greatly. Glucoamylase occurs almost exclusively in fungi, pullulanase in bacteria, and α-D-glucosidase and isoamylase are produced by both fungi and bacteria. Although these enzymes are used by the biochemist as research tools and their mode of action is clearly relevant to the subject of this chapter, only the starch-degrading enzymes found in plants are described here. The reader is referred to the excellent review of Galliard (1987) for information on the fungal and bacterial amylases.

Plants have a number of enzymes that can contribute to starch breakdown and they have been studied in some detail. Endoamylases such as α-amylase (EC 3.2.1.1) can cleave hydrolytically 1,4-α-glucosidic bonds, producing a mixture of linear and branched oligosaccharides, and eventually maltotriose, maltose, glucose, and a range of branched α-limit dextrins. Starch and other α-glucans can be hydrolyzed by β-amylase (EC 3.2.1.2), which catalyzes the removal of successive maltose units from the nonreducing end of α-glucan chains. The maltooligosaccharides produced by these amylolytic enzymes can be further hydrolyzed to glucose by α-glucosidase (EC 3.2.1.20). Manners (1985) presents a good discussion of the need for extensive purification before characterization of starch degradative enzymes (see also the chapter, "Branching Enzymes," for similar points on the characterization of branching enzymes) and the artifacts encountered by workers who did not follow this rule.

For a glucan to be a substrate for starch phosphorylase, it must be longer than maltotetraose. Shorter oligosaccharides, however, can be used by glucosyltransferases such as D-enzyme (EC 2.4.1.25) in the reaction

$$\alpha\text{-glucan} + \alpha\text{-glucan} \Leftrightarrow \alpha\text{-glucan}_{(m+n-1)} + \text{glucose}$$

These enzymes increase the degree of polymerization of short oligosaccharides, converting them into suitable substrates for starch phosphorylase.

None of the enzymes discussed so far can hydrolyze the α-1,6 bonds present in starch. These branch points are cleaved hydrolytically by the debranching enzyme (EC 3.2.1.41), which releases linear oligosaccharides for further metabolism by amylases or phosphorylase.

II. DEBRANCHING ENZYMES

The enzymes that hydrolyze $(1 \rightarrow 6)$-α-D-glucosidic linkages in starch and glycogen and in related α- and β-dextrins are called debranching enzymes. The nomenclature for this type of enzyme was confusing, but has been clarified.

The latest International Union Biochemicals report includes an enzyme 3.2.1.41 α-dextrin endo-1,6-α-glucosidase, other names limit dextrinase, amylopectin 6-glucanohydrolase, pullulanase.

Dohlert and Knutson (1991) and D. J. Manners (personal communication) reported that extracts of sugary maize contain a mixture of limit dextrinase and isoamylase. However, James *et al.* (1995) reported that *su* 1 codes for the isoamylase.

The plant and bacterial enzymes capable of hydrolyzing pullulan do not have identical specificities. In particular, the plant enzymes have little or no action on glycogen and phytoglycogen under conditions in which they readily hydrolyze amylopectin and its β-dextrin. To stress this difference (the bacterial enzymes are capable of degrading both glycogen and phytoglycogen), Manners (1997) recommended different nomenclature for bacterial enzymes, to be called *pullulanase,* and the plant enzymes, to be called *limit dextrinases.*

Manners' (1997) classification is based on the specificity for the substrate [e.g., how readily they hydrolyze glycogen or pullulan—a glucan synthesized by *Pullularia pullulans* (*Aureobasidium pullulans*)—consisting essentially of a linear chain of $(1 \rightarrow 6)$-linked α-maltotriose residues]. According to this classification, higher plant debranching enzymes are called either *limit dextrinases,* which hydrolyze amylopectin, pullulan, and α-dextrins, but do not hydrolyze glycogen, or *isoamylases,* which act on amylopectin, glycogen, and some α-dextrins, but not on pullulan. A third debranching activity, R-enzyme, was discovered independently of limit dextrinases but was later found to be the same enzyme. Hizukuri (1995) classifies them (according to their mode of action) direct and indirect debranching enzymes. The enzymes belong to the first-group if they hydrolyze $(1 \rightarrow 6)\alpha$-branch linkages in one step (e.g., isoamylase and pullulanase). The second group has an indirect action (e.g., an enzyme from rabbit muscle, the amylo 1,6-glucosidase 4-α-D-glucanotransferase, hydrolyzes only a single glucosyl side-chain residue after the other residues of the chain have been transferred to other chains); it also has activity as a 1,4-α-glucanotransferase.

James *et al.* (1995) cloned the *Sugary* 1 (*Su* 1) gene of maize; when this gene is disrupted, the endosperm has reduced amounts of starch and substantial amounts of a water-soluble polysaccharide reminiscent of glycogen (hence its name, *phytoglycogen*). The *Su* 1 cDNA was expressed in *Escherichia coli,* where it displayed debranching activity. The deduced amino-acid sequence of the product of *Su* 1 had strong homology with the bacterial isoamylases. Thus, preliminary evidence suggests that this enzyme could participate in the debranching of a amylopectin precursor, leading to the formation of an amylopectin capable of crystallizing and forming the water-insoluble starch granule (see the chapter, "Open Questions and Hypotheses in Starch").

It is likely that, as research on the structure and function of the plant debranching enzymes progresses, and as the enzymes are compared with their bacterial and fungal counterparts, classification will be based on both mode of action and structure.

III. THE PATHWAY OF STARCH DEGRADATION IN PLANTS

The enzymes capable of degrading starch have been discussed, but the steps required to convert the intact starch granule into soluble maltooligosaccharides have not been addressed yet. Which of the enzyme(s) named in the preceding text are actually involved in the initial degradation of the intact and insoluble starch granule? Is degradation initiated by a particular α-amylase or by one of the phosphorylases? Although some thought that only endoamylases were capable of catalyzing the initial steps of starch degradation, it is known that phosphorylase from pea chloroplasts can release labeled glucose 1-phosphate from ^{14}C-labeled starch granules (Kruger and ap Rees, 1983). It is worth noting that because starch is accumulated in many different tissues, some variation in the pathway of degradation should be expected, if only to accommodate the different metabolic demands of those tissues. For example, reserve starch is compartmentalized differently within legume cotyledon and cereal endosperm. Within the living cotyledon cells, the amyloplast membranes disintegrate during seed maturation and the granules are exposed to the cytosolic enzymes. In the endosperm of cereal seeds, the cells die early and, at maturity, the starch granules are embedded in a matrix of storage protein and are surrounded by a net of dead cell walls. During germination of cereal grains, starch breakdown is associated with the destruction of the endosperm, which softens and eventually liquifies. Starch degradation in these circumstances can be considered to be extracellular, and it seems to occur via a hydrolytic route rather than by phosphorolysis (Boyer, 1985b). The glucose released by the combined action of α-amylases, debranching enzymes, and α-glucosidases is absorbed by the scutellum, converted into sucrose, and transported to the embryo. In other seeds, germination occurs without the destruction of storage tissue, although it is likely that amyloplast membranes are broken down, making starch accessible to cytosolic enzymes. In one of these cases, which is similar to that in pea cotyledons, starch breakdown may be phosphorolytic, but in soybeans and lentils it seems that α-amylolysis may be the major route of starch breakdown. The role of β-amylase in the many seeds is uncertain and, although it may contribute to the degradation of oligosaccharides released by α-amylase, β-amylase does not seem to be essential for starch breakdown (ap Rees, 1988).

The onset of the dark period induces an amylolytic enzyme in the leaves of *Arabidopsis thaliana* plants grown under a light–dark cycle (Kakefuda and Preiss, 1997). Zymograms, obtained from samples harvested over a 24-hour period, indicated that the activity of this hydrolase is induced by the onset of darkness, is higher after 1 hour, and then rapidly disappears from the soluble protein fraction. The peak of activity of this amylase coincides with a decrease in starch content in the leaf. The dark-induced amylase was separated from other, noninducible hydrolases, using polyethylene glycol precipitation, DE-52 ion exchange, Bio-Gel A-0.5M gel filtration, and amylose affinity chromatography. The partially purified enzyme was characterized using the release of Remazol brilliant blue dye from starch azure, hydrolysis of beta-limit dextrin and periodated amylose, and the results obtained indicate that it is an endo-(α)amylase. Activity was inhibited by dialysis of the enzyme preparation against buffers containing EDTA or EGTA, and calcium protected the activity from this inhibition. This amylase differs from other amylolytic enzymes in that it has a relatively high pH optimum for activity (pH 7.0).

Although native starch granules from storage organs seem to be first eroded by amylases before other enzymes can further hydrolyze it, the chloroplastic phosphorylase can release labeled glucose 1-phosphate from ^{14}C-labeled starch granules, at least in pea leaves (Kruger and ap Rees, 1983).

IV. STARCH DEGRADATIVE ENZYMES LOCATED OUTSIDE THE CHLOROPLAST: POSSIBLE FUNCTION

In germinating seeds and tubers, starch degradation is essentially an irreversible process and continues until all the starch is metabolized. This is not so in young and mature leaves, in which starch is a temporary reserve and cellular and organelle integrity is maintained. In the case of young and mature leaves, at least the initial stages of starch breakdown occur inside the chloroplast, and breakdown seems to be mobilized by the joint action of endoamylases and the chloroplastic phosphorylase (Stitt and Steup, 1984).

Incubation of spinach leaf starch granules with extrachloroplastic phosphorylase resulted in the formation of glucose 1-phosphate (Steup *et al.*, 1983). Hammond and Preiss (1983) reported a large increase in cytosolic phosphorylase from spinach leaf in a time course that approximates the time of leaf senescence (i.e., when starch chloroplast must be hydrolyzed and exported to active sinks).

The localization of β-amylase has been a matter of discussion for some time. Okamoto and Akazawa (1979) found β-amylase to be associated with starch granules, an obvious location for an enzyme capable of metabolizing starch, but later, cell fractionation (Beck and Ziegler, 1989) indicated that β-amylase was localized in the vacuole. Wang *et al.* (1995), using monoclonal antibodies selected for phloem-specificity, proposed that the β-amylase of *A. thaliana* is localized in the phloem. They attributed the putative localization of the enzyme in the vacuole and its association with starch granules to the similarity in epitopes between β-amylase and phosphorylase (i.e., it was the starch phosphorylase that had been located and had been described in previous publications). Regarding the possible role for the β-amylase, Wang *et al.* (1995) suggested that it might be there to prevent buildup of starch in the sieve elements during sugar translocation. In seeds, it has been suggested that the enzyme does not fulfill any significant metabolic role (mutants deficient in β-amylase can germinate and grow normally), but is rather a storage protein.

V. DIGESTION OF STARCH IN HUMANS

Starch is a major nutrient in human diet. Digestion of starch consists in the breaking up of the glycosidic bonds linking the glucose residues by glycosidases in order to liberate the reducing components; cooking hydrates starch, making its digestion more efficient.

The principal locations for digestion of starch in humans are the mouth, the lumen of the small intestine, and the brush border of the epithelial cells of the intestinal mucosa. Food is masticated in the mouth, forming a bolus ready for swallowing, while the salivary α-amylase attacks the hydrated starch. The enzyme breaks the starch at random intervals, hydrolyzing internal α-1,4 bonds (and not the α-1,6 bonds constituting the branching points). The α-amylase will not break the bonds nearest the nonreducing ends, nor the bonds next to the α-1,6 branch points, and for this reason the product of its action is a mixture of glucose, maltose, and smaller units of the starch molecule called *starch dextrins,* which contain all of the original α-1,6 bonds.

When food that has been thoroughly chewed reaches the stomach, acidity inactivates the salivary α-amylase, but by then the large starch molecules have been reduced from several thousands to a few glucose units, provided the food has been chewed thoroughly. Little hydrolysis of the carbohydrate occurs in the stomach, although sucrose can be broken up to some extent because the linkage between glucose and fructose, a β-D-fructofuranoside bond, is sensitive to acidity ("acid-labile").

As the stomach empties, the hydrochloric acid in the material entering the small intestine is neutralized by secretions from the pancreatic ducts, bile, and pancreatic juice. The digestion of the starch dextrins is continued by the action of the pancreatic α-amylase.

The pancreatic amylase is very similar to the salivary amylase; indeed, there is only a 1% difference in amino acid composition between the two. Pancreatic amylase is secreted in excess relative to starch intake; it is more important that the salivary enzyme be from a digestive point of view, because food generally does not remain in the mouth long enough to be digested thoroughly by salivary α-amylase. The products of the digestion by α-amylase are mainly maltose and maltotriose, and α-limit dextrins containing about eight glucose units with one or more α-1,6 glucosidic bonds.

It is worth noting that ingested cellulose cannot be digested by humans. Although cellulose is also a polymer of glucose, the linkages between glucose residues are by means of $(1 \rightarrow 4)$-β-D-glycosidic linkages (rather than an α-glycosidic linkages as in starch), and there is no human enzyme capable of hydrolyzing them.

Final hydrolysis of di- and oligosaccharides is carried out by surface enzymes of the small intestinal epithelial cells, called the *brush border,* a term that comes from the appearance of the enterocytes, in which the luminal plasma membrane is enlarged by a regular array of projections called *microvilli.* The enzymes are not secreted into the lumen, but are embedded in the cell membrane, many of these enzymes can protrude into the intestinal lumen up to 10 μm, as they are attached to the plasma membrane by an anchoring polypeptide that has no role by itself in the hydrolysis.

The saccharidases present in the surface of the small intestine relevant to the digestion of starch and its components are:

1. exo-1,4-α-glucosidase (glucoamylase), specific for α-(1,4)glucose bonds

2. oligo-1,6-glucosidase (isomaltase), which acts on α-(1-6)glucose bonds and breaks down isomaltose and α-dextrins
3. α-glucosidase (maltase), which acts specifically on α(1-4)glucose bonds, breaking down maltose and maltotriose

Most of the surface oligosaccharidases are exoenzymes, which clip off one monosaccharide at a time from the nonreducing end. The capacity of the α-glucosidases normally is much greater than that needed for completion of starch digestion. Absorption of the glucose resulting from digestion in the intestinal lumen into the cell is mediated by substrate-specific carriers in the plasma membrane. Glucose transport is the rate-limiting step of carbohydrate at the brush border membrane. Conversely, the duodenal amylase content of an adult is capable of hydrolyzing the starch content of a whole meal in a matter of minutes. Eventually, glucose passes into the portal blood system, through which it is transported first to the liver and then to the remainder of the body.

Because amylase activity is not rate limiting, blood glucose and insulin levels in a healthy adult will be similar if he is fed soluble starch, hydrolyzed starch, low molecular weight maltooligosaccharides, or glucose. Different types of starchy foods can, however, differ in the glycemic responses they elicit because of differences in factors such as particle size (determined in part by milling), amylase inhibitors, amylose–amylopectin ratio, starch–lipid interactions, and starch–protein interactions. Processing may have a large effect on starch availability. For example, raw wheat flour gives a very low sugar and insulin blood levels as compared with the same flour after gelatinization of starch by boiling, and roasting of beans increases the systemic response as compared with boiled beans. In other studies, extrusion cooking of whole grain crisp bread increased starch availability more than conventional baking.

Di-, oligo-, and polysaccharides that are not hydrolized by α-amylase and/or intestinal surface enzymes cannot be absorbed, and they reach the lower tract of the intestine, which, from the lower ileum onward, contains bacteria. Because bacteria contain a larger variety of saccharidases than humans, they can use many of the remaining carbohydrates.

VI. MECHANISM OF ACTION OF AMYLASES AND PHOSPHORYLASES

The nomenclature α- and β- for amylases had its origin in the optical properties of the hydrolysis products of starch. Another nomenclature, proposed in the 1950s, had its basis in the mode of action: endo- and exoamylases. Since that time, research on amylases has moved toward an understanding of the detailed mechanism of action and the crystal structure of the enzymes involved (Yamamoto, 1995). This scientific field is active, partly because of the commercial importance of enzymatic starch degradation for the food and brewing industries.

Polysaccharides, such as starch, phytoglycogen, glycogen, and maltodextrin, are degraded by glucosylhydrolases. Despite large variations in the composition of their substrates, all polysaccharide hydrolases are thought to act by a general acid catalysis mechanism, in which two acidic amino-acid residues participate in either a single or double displacement reaction, resulting in the inversion or retention of configuration at the anomeric carbon atom of the glucosidic bond (Henrissat and Bairoch, 1993). As exemplified by glucoamylase, the nature of the polysaccharide binding site is generally well characterized and consists of several subsites. It is assumed that the total affinity of the oligomeric substrate corresponds to the sum of the individual subsite affinities. Cleavage occurs between subsites 1 and 2, where subsite 2 has the highest affinity of the oligomeric substrate.

In contrast to polysaccharide hydrolases, starch and glycogen phosphorylases (E.C. 2.4.1.1) transfer a glucosyl residue to a phosphate group rather than to water, resulting in the phosphorolytic cleavage of an α-1,4-linked glucose unit and the formation of α-D-glucose 1-phosphate (Glc-1-P) according to the equation

$$[\text{glucose}]_n + P_i \leftrightarrow [\text{glucose}]_{n-1} + \text{glucose 1-phosphate}$$

Another major difference between the phosphorylase-catalyzed reaction and the reaction of hydrolases is the free reversibility of the phosphorolytic cleavage. However, under certain conditions, the reaction of the hydrolysis is reversible, and may result in the synthesis of higher oligosaccharides from maltose and related di- and trioligosaccharides (as studied by French and Hehre in the 1960s and 1970s).

The polysaccharide phosphorylases and the other depolymerizing enzymes differ in both mechanism of action and structure, although it is generally accepted that phosphorylases share the general acid catalysis.

FURTHER READINGS

These sources provide additional in-depth coverage of this topic. For complete reference, please see the Reference section at the end of the book.

Asp, N.-G. (1985)
Beck, E., and Ziegler, P. (1989)

Duffus, C. M. (1984)
Gracey, M., Kretchmer, N., and Rossi, E. (1991)
Hopfer, U. (1997)
Manners, D. J. (1985)
Steup, M. (1984; 1990)
Yamamoto, T. (1995)
Ziegler, P. (1995)

INDUSTRIAL APPLICATIONS OF STARCH

Most of the starch produced in the world is used as food, but about one-third of the total production is employed for a variety of industrial purposes that take advantage of starch's unique properties. The ability to induce starch accumulation in plant organs unable to synthesize it, or to increase starch content at will, would be of great nutritional importance, and some progress has already been made in this respect (see the chapter, "Regulation of the Starch Synthesis Pathway: Targets for Biotechnology"). But quality is as important as quantity. The properties of starch isolated from different sources (e.g., the size of the granules, viscosity, degree of branching, gelation properties) vary greatly and affect the digestibility of the starch and its use in food and nonfood products. Much could be gained by manipulating starch quality—for example, by modifying the ratio of amylose to amylopectin. So far this has been achieved to some extent and in some species by genetic improvement, but in the not-too-distant future, it could be better achieved using recombinant DNA and molecular biology techniques.

It is important for the scientist who works with the basics of carbohydrate metabolism to know how the raw material is used by the industry, and for this reason, here we comment briefly on the use and manipulation of starch in the industrial setting.

I. INDUSTRIAL APPLICATIONS OF STARCH

Estimates (Rabinowitch, 1945) of the total amount of carbon fixed in photosynthesis by land plants is on the order of 2×10^{10} tons per year. Three grain crops (wheat, rice, and maize) and three tuber crops (potato, yam, and cassava) provide $\frac{4}{5}$ of the world's food calculated as calories. The amount of starch produced by the edible portions of these crops in one year exceeds 7×10^8 tons, of which about 6×10^8 tons are contributed by cereal grains.

Most of the starch used in the food and beverage industries is in the form of starch hydrolysates (e.g., glucose, maltose, and isoglucose syrups). In syrup production, yield, ease of processing, color, and flavor are the factors that determine the choice of the raw material. The main application for nonhydrolyzed starch is as a thickener, and for this use the rheologic characteristics of starch pastes are important. These characteristics depend on the botanical source of the starch and include temperature of gelatinization, whether on gelatinization they produce a gel or a sol, whether the gel is opaque or clear, and the gel thickness.

About one-third of the total starch production is used for a variety of industrial purposes that take advantage of its properties (e.g., sizing of paper and board; adhesive in the paper, packaging, and textile industries). In the chemical industry, starch is used as a starting material in fermentative processes to produce polyols, acids, amino acids, cyclodextrins, and fructose.

II. MANUFACTURE AND PROPERTIES OF STARCH

Different processes are used in the manufacture of starch, depending on the plant source, but they all essentially involve freeing the starch granules from the other constituents (e.g., fiber, germ, proteins). Purification is usually done by screening, washing, and centrifugal separation, and the starch obtained is then dried. In the case of maize starch, mechanical separation is facilitated by steeping the grain in warm water in the presence of sulfite. Further processing of starch is carried out on starch suspensions because native starch is insoluble in water.

Little is known about the molecular bases for the differences in the behavior of starches from different sources, but some important factors are the proportion of amylose and amylopectin, the size of the amylose and amylopectin molecules, and the presence of phosphate ester groups (as in potato amylopectin). Potato starch has some favorable characteristics (e.g., its high water retention, slow retrogradation, formation of clear pastes, and lack of a distinctive flavor). Starch from a particular source may have shortcomings that make it inappropriate for a particular use: lack of free flow or water repellency, insolubility or failure to swell and develop viscosity in cold water, excess of viscosity after cooking, cohesive or rubbery texture of cooked starch, tendency to break down during extensive cooking, shear at low pH, lack of clarity, or tendency to become opaque and gel when cooked. The solution in these cases is either to change the source of the raw material or to modify the starch by chemical or physical methods. In maize, sorghum, rice, and barley, *waxy* mutants that have starches essentially free of amylose have been discovered. Starches with high levels of amylose are also known, and some lines of corn have been developed in genetic programs (starting with Whistler and Kramer in the 1940s) with starch containing up to 70% amylose (the normal is around 25%); such cereals are called *amylotypes*.

Viscosity and "mouth-feel" of gravies and other prepared foods, and the behavior of baked products, are determined primarily by the behavior of the starch present in those foods. Starch is responsible for the viscosity and mouth-feel of gravies and puddings and the texture of gumdrops and pie fillings. Baking products "set"—that is, there is a temperature at which the dough no longer expands under the gas pressure generated by the increasing temperature—and the changes experienced by the starch in the dough are responsible to a great degree for that setting. Thus, a great deal of attention has been devoted to the behavior of starch in water. Dried starch granules swell when they are suspended in water; starch can hold up to approximately 30% of its dry weight in moisture. If the temperature is increased slowly, the granules continue to expand slowly. At that stage, they still maintain their crystalline and birefringent properties, and the temperature-induced swelling is still reversible. However, if the temperature is further increased, irreversible swelling occurs, and the granules lose their birefringence. This phenomenon, called *gelatinization* occurs over a range of 5 to 10°C and can be tracked in a number of ways. The optical properties can be examined using a microscope equipped with a hot stage. As temperature is increased, viscosity can be followed using an amylograph (Fig. 1), and as the swelling progresses, so does viscosity, which is measured in Brabender units. Then, as temperature is increased further, the swollen granule breaks down, causing a decrease in viscosity. Starch gelatinization is defined as the loss of birefringence, and this change coincides with a large increase in viscosity. An amylograph measures the relative viscosity of a system as it is heated at a constant rate (1.5°C/min). With continued heating, the starch granule becomes distorted, and soluble starch is released into the solution. The viscosity changes shown by the starch as temperature increases after gelatinization (i.e., after the loss of birefringence) are called *pasting behavior* and vary with the source of the starch and with the amylose content. Heating in the amylograph is stopped at 95°C, the temperature is left constant at that value for some time, and the starch is said to be cooked. During "cooking," the viscosity drops if the system is stirred, because the starch molecules orient themselves in the direction of stirring, a phenomenon called "shear thinning." Another property varies with the source of the starch: The more soluble the starch, the more it "thins" (the more the viscosity drops) on stirring. After the period of time at constant temperature, cooling of the system leads to a rapid increase in viscosity, the "setback," which increases as the hydrogen bonding between the starch molecules increases. This paste

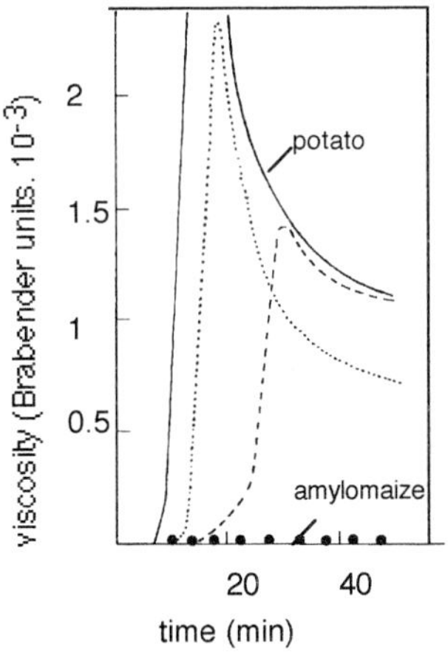

FIG. 1. Pasting behavior of different starches as measured using a Brabender amylograph. Viscosity is measured at varying temperatures; a standard program starts at 50°C, then heats at 1.5°C/min until a temperature of 95°C is reached, and then the temperature is kept constant at 95°C for 30 min. The sources of the different starches is as follows: ——, potato; ······, waxy maize; - - -, maize; ● ●, amylomaize. The first rise in viscosity is linked to the initial swelling of the starch granule, and is followed by a breakdown of swollen granules and a consequent decrease in viscosity. On gelatinization, potato starch swells greatly and then bursts open; cereal starches swell, but tend not to burst, and amylomaize starch swells very little. After Banks and Muir (1980).

is a gel holding large quantities of water. As the paste is cooled, the interactions increase and the gel becomes firmer. If the gel is frozen and thawed, or when it ages, interactions among the starch molecules increase and those with the water decrease, resulting in the loss of water by the gel, which is a phenomenon called *syneresis.* Longer storage leads to even stronger interaction among the starch molecules and eventually to the formation of crystals, a process called *retrogradation.* Crystallization changes the refractive index and the aged gel is more opaque, rigid, and rubbery. Retrogradation is thought to be responsible, at least in part, for the staling of bread, the setback of starch gels, and the skinning of pastes. Waxy starches do not retrograde.

The heat of gelatinization, which can be measured using differential scanning calorimetry, seems to be related directly to the amylopectin content, indicating that amylopectin (and not amylose) is the component responsible for the crystalline structure of starch (see the chapter, "Physicochemical Structure of the Starch Granule").

III. PHYSICAL ANALYSIS OF STARCH AND DERIVATIVES IN THE INDUSTRIAL SETTING

X-ray analysis can be used to differentiate between native starches or to detect changes brought about by physical or chemical treatment of granular starch. Cereal starches give the A pattern, with the exception of amylomaizes, which have an amylose content greater than 40 or 45%. Starch precipitated from pastes by evaporation exhibit different patterns depending on the temperature of evaporation, B if below 50°C, B or C if above 50°C.

Light microscopy supplies information on the dimensions and shape of the granule, and helps to identify the botanical source of the starch and to monitor processing conditions and modifications. The progress of gelatinization can be followed using a polarizing microscope to observe the disappearance of birefringence using a Kofler hot stage, or by using Congo red, which stains gelatinized or broken granules but not native ones. Iodine staining is used to determine the degree of contamination of waxy starch (which stains red) with normal blue-staining granules. Granule size range and diameters can also be measured using a Coulter particle size counter.

The Brabender amylograph is a rotational instrument that permits the measurement of viscosity while cooking and cooling the starch paste. An aqueous dispersion of starch (10% w/w dry basis) is heated to 95°C at a rate of 1.5°C/min. Viscosity is recorded continuously during heating to give the hot-paste viscosity measurements. The peak viscosity indicates the highest viscosity that the industrial user will encounter on preparing a usable paste. The viscosity on reaching 95°C, relative to the peak viscosity, reflects the ease of cooking the starch. After cooking the paste for some time (e.g., 1 hour), the viscosity indicates the stability or degree of breakdown of the paste. Samples in the Brabender amylograph can later be cooled to approximately 50°C for the assessment of the "setback" characteristics (i.e., increased viscosity on cooling). The gels can also be prepared using a viscometer; the gels are cast in molds and, after storage at 20°C for 24 hours, the gel strength is measured as the energy required to compress the gel and the resistance of the gels to rupture.

Polarimetry takes advantage of the high specific rotation of starch, about +200, and is used mainly to measure concentration. It is affected by molecular structure, solvent, and hydrolytic degradation.

Differential scanning calorimetry measures heat flow as a function of temperature. When starch is heated in the presence of excess water, a sharp peak (an endotherm) is obtained, which is caused by the disordering of amylopectin during the gelatinization. The starches that contain lipid give a second endotherm due to the dissociation of an amylose–lipid complex.

IV. CHEMICAL MODIFICATION OF STARCH

Lintner and Naegeli used acid treatment of starch in the late 1800s and, for this reason, acid-treated starches are called *lintnerized* or *Naegeli starches.* In this method, a concentrated starch slurry is heated in 1 to 3% HCl at about 50°C for 12–14 hours, leading to partial acid hydrolysis of the glucosidic bonds in the amorphous portion of the starch. The crystalline areas are not freely accessible to the acid and remain intact. After neutralization and recovery by filtration, the modified starch, with a lower molecular weight but with its crystalline structure intact, displays different characteristics on heating it in water. The granules fragment more and swell less, and the temperature range of gelatinization increases and, on gelatinization, the starch becomes soluble. The resulting paste is less viscous and the gels are more rigid. Acid-modified starches are used in jelly beans and other gum confectioneries.

Crosslinking of starch molecules to make larger ones is performed by forming a diester with phosphoric acid or an ether bond with epichlorohydrin. In a large molecule such as amylopectin, crosslinking can also occur internally. A high degree of crosslinking increases the gelatinization temperature to such an extent that starch can be boiled in water or sterilized in an autoclave and will not gelatinize. A lower degree of crosslinking (which is measured by "degree of substitution") does not affect gelatinization, but strongly affects pasting properties (e.g., viscosity on pasting is decreased and shear thinning decreases—viscosity is not decreased much by shear and pumping) and allows gelatinization in an acidic medium such as that of a cherry pie. Crosslinking slows retrogradation and decreases the changes suffered by the gel on freezing and thawing.

Starch molecules that are modified by forming monoesters of phosphoric acid repel one another because of the added charge, leading to higher swelling and solubilization during gelatinization, and to less interaction during pasting. The resulting starch paste has a higher viscosity, lower resistance to shear thinning, and lower retrogradation and opaqueness. Depending on the degree of substitution, gelatinization can even occur at room temperature, and these starches are useful in the making of instant puddings. Oxidized starches, obtained by treatment with hypochlorite, give better adhesion to meat products but are generally used for nonfood purposes such as the manufacture of paper.

Other useful modifications include acetylation, hydroxyethylation, hydroxypropylation, cationic starches, succinate, and substituted succinate derivatives of starch. In starch-grafted copolymers, free radicals are initiated on the starch that then act as macroinitiators for the acryl or vinyl monomers. These products are used in firefighting fluids, electrolyte solutions for alkaline batteries, wound dressing, and so on.

V. CONVERSION OF STARCH INTO SWEETENERS

In the industry, the degree of hydrolysis is measured as dextrose (glucose) equivalents, or DE, which is a measure of the reducing power of the mixture relative to the number of glucose residues. Simple acid hydrolysis of starch leads to the formation of glucose syrup, because both the α-1,4 and the α-1,6 bonds are susceptible to the acid. However, after a certain DE is obtained, side reactions can occur that can affect the quality of the syrup obtained. For this reason, acid treatment is often used for thinning (i.e., for partial hydrolysis rather than a total degradation). The resulting dextrins can be used directly for increasing the viscosity of a number of products, but for the production of sweeteners, they must be further degraded using enzymes such as α-amylase, β-amylase, and/or glucoamylase. If a very high degree of sweetness is desired, glucose can be converted into fructose by the action of glucose isomerase.

VI. BIODEGRADABLE POLYMERS

Starch and starch derivatives can improve the biodegradability of a polymer as a whole when incorporated as fillers or functional additives (Booma *et al.,* 1994). Although starch-based polymers have also been developed, research continues with the following aims: to increase the speed and degree of degradability, to increase compatibility with inks, to decrease toxicity and hygroscopy, and to improve mechanical and rheologic properties. High amylose starch can be processed by extrusion or injection molding under highly controlled environmental conditions to give biodegradable thermoplastic materials (Fanta and Doane, 1986; Booma *et al.,* 1994). All industrial applications require some modification of the starch (e.g., the addition of plasticizers, chemical bonding, addition of high cost ethylene–acrylic acid copolymers). Although a variety of starches are being used for the manufacture of biodegradable polymers, it is likely that starches from some plants may be more suitable than others for a particular purpose. Indeed, a number of specialty starches (such as amylose and waxy starch) are being increasingly recognized for their specific properties and biodegradability, and there is certainly room for other novel products. Thus, a greater understanding of the critical components of the plant starch biosynthetic machinery could have a major impact on agriculture and industry.

FURTHER READINGS

These sources provide additional in-depth coverage of this topic. For complete reference, please see the Reference section at the end of the book.

Banks, W., and Muir, D. D. (1980)
Farris, P. (1984)
Galliard, T. (1987)
Hoseney, R. C. (1990)
MacMasters, M. M. (1964)
Scofield, J. D., and Greenwell, P. (1987)
Whistler, R. L. (1984)
Wurzburg, O. B. (1986)

REFERENCES

Aebi, M., Hornig, H., and Weissman, C. 1987. 5′ Cleavage site in eucaryotic pre-mRNA splicing is determined by the overall 5′ splice region, not by the conserved 5″ GU. *Cell* **50,** 237–246.

Ainsworth, C., Travis, M., and Clark, J. 1993. Isolation and analysis of a cDNA clone encoding the small subunit of ADP-glucose pyrophosphorylase from wheat. *Plant Mol. Biol.,* **23,** 23–33.

Akazawa, T., Pozueta-Romero, J., and Ardila, F. 1991. New aspects of starch biosynthesis in the plant cell. *In* "Compartmentation and Metabolic Regulation" (A. H. C. Huang and L. Taiz, eds.), pp. 74–85. American Society of Plant Physiologists, Rockville, MD.

Anderson, J. M., Hnilo, J., Larson, R., Okita, T. W., Morell, M., and Preiss, J. 1989. The encoded primary sequence of a rice seed ADP-glucose pyrophosphorylase subunit and its homology to the bacterial enzyme. *J. Biol. Chem.* **264,** 12238–12242.

Anderson, J. M., Larsen, R., Laudencia, D., Kim, W. T., Morrow, Okita, T. W., and Preiss, J. 1991. Molecular characterization of the gene encoding a rice endosperm-specific ADP-glucose pyrophosphorylase subunit and its developmental pattern of transcription. *Gene* **97,** 199–205.

Anderson, J. M., Okita, T. W., and Preiss, J. 1990. Enhancing carbon flow into starch: the role of ADPglucose pyrophosphorylase. *In* "Molecular and Cellular Biology of the Potato" (W. C. Park and M. E. Vayda, eds.), pp. 159–180. C. A. B. International, Wallinford, United Kingdom.

ap Rees, T. 1987. Compartmentation of plant metabolism. *In* "The Biochemistry of Plants" (D. D. Davies, ed.), Vol. 12, pp. 87–115. Academic Press, New York.

ap Rees, T. 1988. Hexose phosphate metabolism by nonphotosynthetic tissues of higher plants. *Biochem. Plants* **14,** 1–33.

ap Rees, T. 1995. Where do plants make ADP-Glc? *In* "Sucrose Metabolism, Biochemistry, Physiology and Molecular Biology" (H. Pontis, G. Salerno, and E. Echeverria, eds.), pp. 143–155. ASPP, Rockville, MD.

ap Rees, T., and Entwistle, G. 1989. Entry in the amyloplasts of carbon for starch synthesis. *In* "Physiology, Biochemistry and Genetics of Non-Green Plastids" (C. D. Boyer, J. C. Shannon, R. C. Hardison, eds.), pp. 49–62. American Society of Plant Physiologists, Rockville, MD.

ap Rees, T., Leja, M., Macdonald, F. D., and Green, J. H. 1984. Nucleotide sugars and starch synthesis in spadix of *Arum maculatum* and suspension cultures of *Glycine max. Phytochemistry* **23,** 2463–2468.

Archbold, H. K. 1940. Fructosans in the monocotyledons: A review. *New Phytol.* **39,** 185–219.

Ardila, F., and Tandecarz, J. S. 1992. Potato tuber UDP-glucose: Protein transglucosylase catalyses its own glucosylation. *Plant Physiol.* **99,** 1342–1347.

Asp, N.-G. 1985. Cereal carbohydrates in human nutrition. *In* "New approaches to Research in Cereal Carbohydrates." (R. D. Hill and L. Munck, eds.) pp. 339–343. Elsevier Sc. Pub. Amsterdam.

Baba, T., Kimura, K., Mizuno, K., Etoh, H., Ishida, Y., Shida, O., and Arai, Y. 1991. Sequence conversion of the catalytic regions of the amylolytic enzymes in maize branching enzyme. *Biochem. Biophys. Res. Commun.* **181,** 87–94.

Baba, T., Nishihara, M., Mizuno, K., Kawasaki, T., Shimada, H., Kobayashi, E., Ohnishi, S., Tanaka, K., and Arai, Y. 1993. Identification, cDNA cloning and gene expression of soluble starch synthase in rice (*Oryza sativa L.*) immature seeds. *Plant Physiol.* **103,** 565–573.

Baba, T., Yoshii, M., and Kainuma, K. 1987. Acceptor molecule of granular-bound starch synthase from sweet potato roots. *Starch/Staërke* **39,** 52–56.

Bae, J. M., Giroux, J., and Hannah, L. C. 1990. Cloning and characterization of the *brittle*-2 gene of maize. *Maydica* **35,** 317–322.

Baecker, P. A., Greenberg, E., and Preiss, J. 1986. Biosynthesis of bacterial glycogen: Primary structure of Escherichia coli 1,4-α-D-glucan: 1,4-α-D-glucan 6-α-D-(glucano)-transferase as deduced from the nucleotide sequence of the glgB gene. *J. Biol. Chem.* **261,** 8738–8743.

Bailey, J. M., and Whelan, W. J. 1961. Physical properties of starch. I. Relationship between iodine stain and chain length. *J. Biol. Chem.* **236,** 969–973.

Ball, K. L., and Preiss, J. 1992. Evidence for an arginine residue at the allosteric sites of spinach leaf ADPglucose pyrophosphorylase. *J. Protein Chem.* **11,** 231–237.

Ball, K. L., and Preiss, J. 1994. Allosteric sites of the large subunit of the spinach leaf adenosine diphosphate glucose pyrophosphorylase. *J. Biol. Chem.* **269,** 24706–24711.

Ball, S., Dirick, L., Decq, A., Martiat, J. C., and Matagne, R. F. 1990. Physiology of starch storage in the monocellular algae *Chlamydomonas reinhardtii. Plant Sci.* **66,** 1–9.

Ball, S., Guan, H.-P., James, M., Myers, A., Keeling, P., Mouille, G., Buléon, A., Colonna, P., and Preiss, J. 1996. From glycogen to amylopectin: A model for the biogenesis of the plant, starch granule. *Cell* **86,** 349–352.

Ball, S., Marianne, T., Dirick, L., Fresnoy, M., Delrue, B., and Decq, A. 1991. A *Chlamydomonas reinhardtii* low-starch mutant is defective for 3-phosphoglycerate activation and orthophosphate inhibition of ADP-glucose pyrophosphorylase. *Planta* **185,** 17–26.

Ballicora, M. A., Fu, Y., Wu, M.-X., Sheng, J., Nesbitt, N. M., and Preiss, J. 1996. Studies on the catalytic and regulatory sites of bacterial and plant ADPGlc pyrophosphorylase. *In* "Regulation and Manipulation of Starch and Sucrose Metabolism in Plants" (Y. Nakamura, ed.), pp. 5–11. National Institute of Agrobiological Resources of Japan, Tokyo.

Ballicora, M. A., Laughlin, M. J., Fu, Y., Okita, T. W., Barry, G. F., and Preiss, J. 1995. ADPglucose from potato tuber: Significance of the N-terminal of the small subunit for catalytic properties and heat stability. *Plant Physiol.* **109,** 245–251.

Banks, W., and Greenwood, C. T. 1975. "Starch and Its Components." Edinburgh Univ. Press, Scotland.

Banks, W., and Muir, D. D. 1980. Structure and chemistry of the starch granule. *In* "The Biochemistry of Plants" (J. Preiss, ed.), Vol 3, pp 321–369. Academic Press, New York.

Barry, G. F., Kishore, G., Stark, D. M., and Zalewsky, J. C. 1994. "Method of improving the quality of stored potatoes." (PCT/US94/05275). World Intellectual Property Organization, International Bureau,

Barton, C., Yang, L., Galvin, M., Sengupta-Gopalan, C., and Borelli, T. 1986. Isolation of the *Shrunken*-2 and *Brittle*-2 genes from maize. *In* "Regulation of Carbon and Nitrogen Reduction and Utilization in Maize" (J. C. Shannon, D. P. Knievel, and C. D. Boyer, eds.), pp. 363–365. American Society of Plant Physiologists, Rockville, MD.

Bassham, J. A., and Calvin, M. 1957. "The Path of Carbon in Photosynthesis." Prentice-Hall, Englewood Cliffs, NJ.

Bassham, J. A., and Krause, G. H. 1969. Free energy changes and metabolic regulation in steady-state photosynthetic carbon reduction. *Biochim. Biophys. Acta* **153,** 211–218.

Batz, O., Maus, U., Henrichs, G., Scheibe, R., Neuhaus, H.-E. 1994. Glucose and ADPGlc-dependent starch synthesis in isolated cauliflower-bud amyloplasts: Analysis of the interaction of various potential precursors. *Biochim. Biophys. Acta* **1200,** 148–154.

Beck, E., and Ziegler, P. 1989. Biosynthesis and degradation of starch in higher plants. *Annu. Rev. Plant Physiol. Plant Mol. Biol.* **40,** 95–117.

BeMiller, J. N. 1997. Structure of the starch granule. *J. Appl. Glycosci.* **44,** 43–49.

Benson, A. A., Bassham, J. A., Calvin, M., Goodale, T. C., Haas, V. A., and Stepka, W. 1950. The path of carbon in photosynthesis: V. Paper chromatography and radioautography of the products. *J. Am. Chem. Soc.* **72,** 1710–1718.

Benson, A. A., Bassham, J. A., Calvin, M., Hall, A. G., Hirsch, H. E., Kawaguchi, S., Lynch, V., and Tolbert, N. E. 1952. The path of carbon in photosynthesis. XV. Ribulose and sedoheptulose. *J. Biol. Chem.* **196,** 703–716.

Benson, A. A., and Calvin, M. 1950. Carbon dioxide fixation by green plants. *Ann. Rev. Plant Physiol.* **1,** 25–42.

Bhattacharyya, M. K., Smith, A. M., Noel-Ellis, T. H., Hedley, C., and Martin, C. 1990. The wrinkled-seed character of pea described by Mendel is caused by a transposon-like insertion in a gene encoding starch-branching enzyme. *Cell* **60,** 115–122.

Bhave, M. R., Lawrence, S., Barton, C., and Hannah, L. C. 1990. Identification and molecular characterization of *Shrunken*-2 cDNA clones of maize. *Plant Cell* **2,** 581–588.

Binderup, K. 1997. Site directed mutagenesis studies and crystallization experiments in the *E. coli* branching enzyme. Master's thesis. Århus University, Denmark.

Binderup, K., and Preiss, J. In press. Glu-459 is important in function of *E. coli* branching enzyme. *Biochemistry.*

Bird, I. F., Cornelius, M. J., Keys, A. J., and Whittingham, C. P. 1974. Intracellular site of sucrose synthesis in leaves. *Phytochem.* **13,** 59–64.

Blennow, A., and Johnson, G. 1991. Isolation of Q-enzyme with Mr 103,000 from potato tubers. *Phytochem.* **30,** 437–444.

Bocca, S. N., Rothschild, A., and Tandecarz, J. S. 1997. Initiation of starch biosynthesis: Purification and characterization of UDP-glucose. Protein transglucosylase from potato tubers. *Plant Physiol. Biochem.* **35,** 305–312.

Booma, M., Selke, S. E., and Giacin, J. R. 1994. Biodegradable plastics. *J. Elastom. Plastics* **26,** 104–142.

Borchert, S., Grosse, H., and Heldt, H. W. 1989. Specific transport of inorganic phosphate, glucose 6-phosphate, dihydroxyacetone phosphate and 3-phospho-glycerate into amyloplasts from pea roots. *Fed. Eur. Biochem. Soc.* **253,** 183–186.

Borchert, S., Harbroth, J, Schünemann, D., Hoferichter, P., and Heldt, H. W. 1993. Studies of the enzymic capacities and transport properties of pea root plastids. *Plant Physiol.* **101,** 303–312.

Borovsky, D., Smith, E. E., and Whelan, W. J. 1975. Purification and properties of potato 1,4-α-D-glucan: 1,4-α-D-glucan, 6-α-(1,4-α-D-glucano)-transferase. Evidence a dual catalytic function in amylose branching enzyme. *Eur. J. Biochem.* **59,** 615–625.

Bourne, E. J., and Peat, S. 1945. The enzymic synthesis and degradation of starch. I. The synthesis of amylopectin. *J. Chem. Soc.* 877–882.

Boyer, C., and Preiss, J. 1979. Properties of citrate stimulated starch synthesis catalyzed by starch synthase I of developing maize kernels. *Plant Physiol.* **64,** 1039–1042.

Boyer, C. D. 1985b. Synthesis and breakdown of starch. *In* "Biochemical Basis of Plant Breeding" (C. A. Neyra, ed.), Vol. 1, pp. 133–153. CRC Press, Boca Raton, FL.

Boyer, C. D., and Preiss, J. 1978a. Multiple forms of $(1\rightarrow4)$-α-D-glucan 6-glycosyl transferase from developing *Zea mays* L. kernels. *Carbohydr. Res.* **61,** 321–334.

Boyer, C. D., and Preiss, J. 1978b. Multiple forms of starch branching enzyme of maize: Evidence for independent genetic control. *Biochem. Biophys. Res. Commun.* **80,** 169–175.

Boyer, C. D., and Preiss, J. 1981. Evidence for independent genetic control of the multiple forms of maize endosperm branching enzymes and starch synthases. *Plant Physiol.* **67,** 1141–1145.

Boyer, C. D., Shannon, J. C., and Hardison, R. C., eds. 1989. "Physiology, Biochemistry and Genetics of Nongreen Plastids." American Society of Plant Physiologists, Rockville, MD.

Brangeon, J., Reyss, A., and Prioul, J.-L. 1997. In situ detection of ADPglucose pyrophosphorylase expression during maize endosperm development. *Plant Physiol. Biochem.* **35,** 847–858.

Breathnach, R., and Chambon, P. 1981. Organization and expression of eucaryotic split genes coding for proteins. *Ann. Rev. Biochem.* **50,** 349–383.

Brinson, K., and Dey, P. M. 1985. Polysaccharides containing xylose, arabinose, and galactose in higher plants. *In* "Biochemistry of Storage Carbohydrates in Green Plants" (P. M. Dey and R. A. Dixon, eds.), pp. 349–371. Academic Press, New York.

Brown, J. W. S. 1986. A catalogue of splice junction and putative branch-point sequences from plant introns. *Nucl. Acid Res.* **14,** 9549–9559.

Bucke, C., and Oliver, I. R. 1975. Location of enzymes metabolising sucrose and starch in the grasses *Pennisetum purpureum* and *Muhlenbergia montana. Planta* **122,** 45–52.

Burns, J. A., Cornish-Bowden, A., Groen, A. K., Heinrich, R., Kasser, H., Porteous, J. W., Rapoport, S. M., Rapoport, T. A., Stucki, J. W., Tager, J. M., Wanders, R. J. A., and Westerhoff, H. V. 1985. Control analysis of metabolic systems. *TIBS.* **10,** 16.

Callis, J., Fromm, M., and Walbot, V. 1987. Introns increase gene expression in cultured maize cells. *Genes and Devel.* **1,** 1183–1200.

Campbell, D. G., and Cohen, P. 1989. The amino acid sequence of rabbit skeletal muscle glycogenin. *Eur. J. Biochem.* **185,** 119–125.

Cao, H., and Preiss, J. 1996. Evidence for essential arginine residues at tre active site of maize branching enzymes. *J. Protein Chem.* **15,** 291–304.

Cao, H., Sullivan, T. D., Boyer, C. D., and Shannon, J. C. 1995. Bt1, a structural gene for the major 39-44 kDa amyloplast membrane polypeptides. *Physiol. Plant.* **95,** 176–186.

Cardini, C. E., Leloir, L. F., and Chiriboga, J. 1955. The biosynthesis of sucrose. *J. Biol. Chem.* **214,** 149–155.

Carter, J., and Smith, E. E. 1978. Actions of glycogen synthase and phosphorylase of rabbit-skeletal muscle on modified glycogens. *Carbohydr. Res.* **61,** 395–406.

Caspar, C., Huber, S. C., and Somerville, C. 1986. Alterations in growth, photosynthesis and respiration in a starch mutant of *Arabidopsis thaliana* (L.), Heynh deficient in chloroplast phosphoglucomutase activity. *Plant Physiol.* **79,** 1–7.

Caspar, T. 1994. Genetic dissection of the biosynthesis, degradation and biological functions of starch. *In* "Arabidopsis" (Monograph 27) (E. M. Meyerowitz and C. Sommerville, eds.), pp. 913–936. Cold Spring Harbor Lab. Press, Coldspring Harbor, New York.

Caspar, T., and Pickard, B. G. 1989. Gravitropism in a starchless mutant of *Arabidopsis* implications for the starch-statolith theory of gravity sensing. *Planta* **177,** 185–197.

Charng, Y.-Y., Iglesias, A. A., and Preiss, J. 1994. Structure–function relationships of cyanobacterial ADP-glucose pyrophosphorylase: Site-directed mutagenesis and chemical modification of the activator-binding sites of ADP-glucose pyrophosphorylase from *Anabaena* PCC 7120. *J. Biol. Chem.* **269,** 24107–24113.

Charng, Y.-Y., Kakefuda, G., Iglesisas, A. A., Buikema, W. J., and Preiss, J. 1992. Molecular cloning and expression of the gene encoding ADP-glucose pyrophosphorylase from cyanobacterium *Anabaena sp.* strain PCC 7120. *Plant Mol. Biol.* **20,** 37–47.

Charng, Y.-Y., Sheng, J., and Preiss, J. 1995. Mutagenesis of an amino acid residue in the activator-binding site of ADP-glucose pyrophosphorylase causes alteration in activator specificity. *Arch. Biochem. Biophys.* **318,** 476–480.

Chen, T. M., Dittrich, P., Campbell, W. H., and Black, C. C. 1974. Metabolism of epidermal tissues, mesophyll cells and bundle sheath strands resolved from mature nutsledge leaves. *Arch. Biochem. Biophys.* **163,** 246–262.

Ching, T. M., Poklemba, C. J., and Metzger, R. J. 1983. Starch synthesis in shriveled and plump *Triticale* seeds. *Plant Physiol.* **73,** 652–657.

Clark, J. R., Robertson, M., and Ainsworth, C. C. 1991. Nucleotide sequence of a wheat (*Triticum aestivum* L.) cDNA clone encoding the *waxy* protein. *Plant Mol. Biol.* **16,** 1099–1101.

Coates, S. A., and ap Rees, T. 1994. Metabolism of glucose monophosphates by leucoplasts and amyloplasts from soybean suspension cultures. *Phytochemistry* **35,** 881–883.

Cohn, M. 1961. *In* "The Enzymes" (P. D. Boyer, H. Lardy, and K. Myrback, eds.) 2nd. ed., Vol 5, pp. 179–206. Academic Press. New York.

Dang, P. L., and Boyer, C. D. 1988. Maize leaf and kernel starch synthases and branching enzymes. *Phytochemistry* **27,** 1255–1259.

de Fekete, M. A. R., and Cardini, C. E. 1964. Mechanism of glucose transfer from sucrose into the starch granule of sweet corn. *Arch. Biochem. Biophys.* **104,** 173–184.

Dedio, W., Simmonds, D. H., Hill, R. D., and Shealy, H. 1975. Distribution of alpha-amylase in the triticale kernel during development. *Can. J. Plant Sci.* **55,** 29–36.

Delrue, B., Fontaine, T., Routier, F., Decq. A., Wieruszeski, J.-M., Van Den Koornhuyse, N., Maddelein, M.-L., Fournet, B., and Ball, S. 1992. Waxy *Chlamydomonas reinhardtii:* Monocellular algal mutants defective in amylose synthesis and granule-bound starch synthase activity accumulate a structurally modified amylopectin. *J. Bacteriol.* **174,** 3612–3620.

Dennis, D. T., Heckman, W. E., Thomson, A., Ireland, R. J., Botha, F. C., and Kruger, N. J. 1985. *In* "Regulation of Carbon Partitioning in Photosynthetic Tissue" (R. Heath, and J. Preiss, eds.), pp. 127–146. American Society of Plant Physiologist Rockville, MD.

Denyer, K., Dunlap, F., Thorbjøtornsen, T., Keeling, P., and Smith, A. M. 1996. The major form of ADPglucose pyrophosphorylase in maize endosperm is extra-plastidial. *Plant Physiol.* **112,** 779–785.

Denyer, K., Sidebottom, C., Hylton, C. M., and Smith, A. M. 1993. Soluble isoforms of starch synthase and starch-branching enzyme also occur within starch granules in developing pea embryos. *Plant J.* **4,** 191–196.

Denyer, C., and Smith, A. 1988. The capacity of plastids from developing pea cotyledons to synthesize acetyl CoA. *Planta* **173,** 172–182.

Dickinson, D. B., and Preiss, J. 1969a. ADPglucose pyrophosphorylase from maize endosperm. *Arch. Biochem. Biophys.* **130,** 119–128.

Dickinson, D. B., and Preiss, J. 1969b. Presence of ADP-glucose pyrophosphorylase in *Shrunken*-2 and *Brittle*-2 mutants of maize endosperm. *Plant Physiol.* **44,** 1058–1062.

Dohlert, D. C., and Knutson, C. A. 1991. Two classes of starch debranching enzymes from developing maize kernels. *J. Plant Physiol.* **138,** 566–572.

Downey, K., Voellmy, R., Ahmad, F., and Schultz, J. (Eds.) 1983. "Advances in Gene Technology: Molecular Genetics of Plants and Animals." Academic Press New York.

Downton, W. J. S., and Hawker, J. 1973. Enzymes of starch and sucrose metabolism in *Zea mays* leaves. *Phytochem.* **12,** 1551–1556.

Dry, I., Smith, A., Edwards, A., Bhattarcharyya, M., Dunn, P., and Martin, C. 1992. Characterization of cDNAs encoding two isoforms of granule-bound starch synthase which show differential expression in developing storage organs of pea and potato. *Plant J.* **2,** 193–202.

Echeverria, E., and Boyer, C. D. 1986. Localization of starch synthetic and degradative enzymes in maize leaves. *Am. J. Bot.* **73,** 167–171.

Echeverria, E., Boyer, C. D., Liu, K.-C. and Shannon, J. C. 1985. Isolation of amyloplasts from developing maize endosperm. *Plant Physiol.* **77,** 513–519.

Echeverria, E., Boyer, C. D., Thomas, P. A., Liu, K.-C., and Shannon, J. C. 1988. Enzyme activities associated with maize kernel amyloplasts. *Plant Physiol.* **86,** 786–792.

Edwards A., Marshall, J., Denyer, K., Sidebottom, C., Visser, R. G., Martin, C., and Smith, A. M. 1996. Evidence that a 77-kilodalton protein from the starch of pea embryos is an isoform of starch synthase that is both soluble and granule-bound. *Plant Physiol.* **112,** 89–97.

Edwards, G., and Walker, D. A. 1983. "C_3, C_4: Mechanisms, Cellular and Environmental Regulation of Photosynthesis." Blackwell Scientific Publishers, Oxford, England.

Edwards, J., Green, J. H., and ap Rees, T. 1988. Activity of branching enzyme as a cardinal feature of the Ra locus in *Pisum sativum. Phytochem.* **27,** 1615–1620.

Entwistle, G., and ap Rees, T. 1988. Enzyme capacities of amyloplasts from wheat (*Triticum aestivum*) endosperm. *Biochem. J.* **255,** 391–396.

Entwistle, G., Tyson, R. H., and ap Rees, T. 1988. Isolation of amyloplasts from wheat endosperm. *Phytochemistry* **25,** 2033–2039.

Erlander, S. 1958. Proposed mechanism for the synthesis of starch from glycogen. *Enzymol.* **19,** 273–283.

Espada, J. 1962. Enzymic synthesis of adenosine diphosphate glucose from glucose 1-phosphate and adenosine triphosphate. *J. Biol. Chem.* **237,** 3577–3581.

Fahrendorf, T., Holtum, J. A. M., Muckerjee, U., and Latzko, E. 1987. Fructose 2,6-biphosphate, carbohydrate partitioning, and crassulacean acid metabolism. *Plant Physiol.* **84,** 182–187.

Fales, F. W. 1980. The linear relationship between iodine staining and average chain length of the unbranched amyloglucans. *Biopolymers* **19,** 1535–1542.

Fanta, G. F., and Doane, W. M. 1986. *In* "Modified Starches: Properties and Uses" (O. B. Wurzburg, ed.) pp. 149–178. CRC Press, Boca Raton, FL.

Farris, P. 1984. Economics and future of the starch industry. *In* "Starch: Chemistry and Technology" (R. L. Whistler, J. N. BeMiller, and E. F. Paschall, eds.), 2nd ed., pp. 11–24. Academic Press, New York.

Fedoroff, N., Wessler, S., and Shure, M. 1983. Isolation of the transposable maize controlling elements *Ac* and *Ds. Cell* **35,** 235–242.

Fekete, M. A. R. de, Leloir, L. F., and Cardini, C. E. 1960. Mechanism of starch biosynthesis. *Nature* (London), **187,** 918–919.

Fekete, M. A. R. de, Leloir, L. F., and Cardini, C. E. 1961. Starch and oligosaccharide synthesis from uridine diphosphate glucose. *J. Biol. Chem.* **236,** 636–641.

Feldman, K. A., and Marks, M. D. 1987. *Mol. Gen. Genet.* 208, 1–9.

Fisher, D. K., Boyer, C. D., and Hannah, L. C. 1993. Starch branching enzyme II from maize endosperm. *Plant Physiol.* **102,** 1053–1054.

Fisher, D. K., Gao, M., Kim, K.-N., Boyer, C. D., and Guiltinan, M. J. 1996a. Allelic analysis of the maize *amylose-extender* locus suggests that independent genes encode starch branching enzymes IIa and IIb. *Plant Physiol.* **110,** 611–619.

Fisher, D. K., Gao, M., Kim, K.-N., Boyer, C. D., and Guiltinan, M. J. 1996b. Two closely related cDNAs encoding starch branching enzyme from *Arabidopsis thaliana. Plant Mol. Biol.* **30,** 97–108.

Fisher, M. B., and Boyer, C. D. 1983. Immunological characterization of starch branching enzymes. *Plant Physiol.* **72,** 813–816.

Fitt, L. E., and Snyder, E. M. 1984. Photomicrographs of starch. *In* "Starch, Chemistry and Technology" (R. L. Whistler, J. N. BeMiller, and E. F. Paschall, eds.), 2nd ed., pp. 675–689. Academic Press, New York.

Fontaine, T., D'Hulst, C., Maddelein, M.-L., Routier, F., Pepin, T. M., Decq, A., Wieruszeski, J.-M., Delrue, B., Van Den Koornhuyse, N., Bossu, J.-P., Fournet, B., and Ball, S. 1993. Toward an understanding of the biogenesis of the starch granule; Evidence that *Chlamydomonas* soluble starch synthase II controls the synthesis of intermediate size glucans of amylopectin. *J. Biol. Chem.* **268,** 16223–16230.

Fox, J., Kawaguchi, K., Greenberg, E., and Preiss, J. 1976. Biosynthesis of bacterial glycogen. Purification and properties of the *Escherichia coli* B ADPglucose: 1,4-α-D-glucan 4-α-glucosyltranferase. *Biochem.* **15,** 849–857.

French, D. 1972. Fine structure of starch and its relationship to the organization of starch granules. *J. Jpn. Soc. Starch Sci.* **19,** 8–25.

French, D. 1984. Organization of starch granules. *In* "Starch, Chemistry and Technology" (R. L. Whistler, J. N. BeMiller, and E. F. Paschall, eds.), 2nd ed., pp. 183–247. Academic Press, New York.

Frydman, R. B., and Cardini, C. E. 1966. Studies on the biosynthesis of starch II. *J. Biol. Chem.* **242,** 312–317.

Furbank, R. T., Stitt, M., and Foyer, C. H. 1985. Intercellular compartmentation of sucrose synthesis in leaves of *Zea mays. Planta.* **164,** 172–178.

Furlong, C. E., and Preiss, J. 1969. Biosynthesis of bacterial glycogen synthesis. VII. Purification and properties of adenosine diphosphoglucose pyrophosphorylase of *Rhodospirillum rubrum. J. Biol. Chem.* **244,** 2539–2548.

Furukawa, K., Tagaya, M., Inouye, M., Preiss, J., and Fukui, T. 1990. Identification of lysine 15 at the active site in *Escherichia coli* glycogen synthase: Conservation of a Lys-X-Gly-Gly sequence in the bacterial and mammalian enzymes. *J. Biol. Chem.* **265,** 2086–2090. 871.

Furukawa, K., Tagaya, M., Tanizawa, K., and Fukui, T. 1993. Role of the conserved Lys-X-Gly-Gly sequence at the ADP-glucose-binding site in *Escherichia coli* glycogen synthase. *J. Biol. Chem.* **268,** 23837–23842.

Furukawa, K., Tagaya, M., Tanizawa, K., and Fukui, T. 1994. Identification of Lys^{277} at the active site of *Escherichia coli* glycogen synthase. *J. Biol. Chem.* **269,** 868–871.

Galliard, T. 1987. "Starch: Properties and Potential." John Wiley and Sons, New York.

Gamanetz, L. V., and Gamburg, K. Z. 1981. The effect of adenine on growth, starch and ADPG adenosine-5′-(α-D-glucopyranosyl)-pyrophosphate and ADPG pyrophosphorylase content in suspension cultures tobacco cells. *Z. Pflanzenphysiol.* **104,** 61–69.

Gardiol, A., and Preiss, J. 1990. *Escherichia coli* E-39 ADPglucose synthetase has different activation kinetics from the wild type allosteric enzyme. *Arch. Biochem. Biophys.* **280,** 175–180.

Geddes, R., Greenwood, C. T., and Mackenzie, S. 1965. Studies on the biosynthesis of starch granules. Part III. The properties of the components of starches from the growing potato tubers. *Carbohyd. Res.* **1,** 71–82.

Gentner, N., Greenberg, E., and Preiss, J. 1969. TPNH and pyridoxal-5′-phosphate: Activators of ADP-glucose pyrophosphorylase of *Escherichia coli* B. *Biochem. Biophys. Res. Commun.* **36,** 373–380.

Ghosh, P., Meyer, C., Remy, E., Peterson, D., and Preiss, J. 1992. Biosynthesis of bacterial glycogen: Cloning, expression and nucleotide sequence of *glg* C Gene from an allosteric mutant of *Escherichia coli* B. *Arch. Biochem. Biophys.* **296,** 122–128.

Ghosh, H. P., and Preiss, J. 1966. Adenosine diphosphate glucose pyrophosphorylase: A regulatory enzyme in the biosynthesis of starch in spinach chloroplasts. *J. Biol. Chem.* **241,** 4491–4505.

Giroux, M. J., Shaw, J., Barry, G., Cobb, B. G., Greene, T., Okita, T., and Hannah, L. C. 1996. A single gene mutation that increases maize seed weight. *Proc. Natl. Acd. Sci. USA,* **93,** 5824–5829.

Goldner, W., and Beevers, H. 1989. Starch synthase and starch branching enzyme from germinating castor bean endosperm. *Phytochemistry* **28,** 1809–1812.

Gough, B. M., Greenwell, P., and Russell, P. L. 1985. On the interaction of sodium dodecyl sulphate with starch granules. *In* "New Approaches to Research on Cereal Carbohydrates" (R. D. Hill and L. Munck, eds.), pp. 99–108. Elsevier Scientific Publishers, Amsterdam, The Netherlands.

Gracey, M., Kretchmer, N., and Rossi, E. 1991. "Sugars in Nutrition." Raven Press, New York.

Greenberg, E., Preiss, J. E., VanBoldrick, M., and Preiss, J. 1983. Biosynthesis of bacterial glycogen: Activator specificity of the ADPglucose pyrophosphorylase of *Rhodopseudomonads. Arch. Biochem. Biophys.* **220,** 594–604.

Greene, T. W., Chantler, S. E., Kahn, M. L., Barry, G. F., Preiss, J., and Okita, T. W. 1996. "Mutagenesis of the potato ADPglucose pyrophosphorylase and characterization of an allosteric mutant defective in 3-phosphoglycerate activation." *Proc. Natl. Acad. Sci.* **93,** 1509–1513.

Greene, T. W., Chantler, S. E., Kahn, M. L., Barry, G. F., Preiss, J., and Okita, T. W. 1996a. *Proc. Natl. Acad. Sci. USA* **93,** 1509–1513.

Greene, T. W., Woodbury, R. L., and Okita, T. O. 1996b. *Plant Physiol.* **112,** 1315–1320.

Greenwell, P., Evers, A. D., Gough, B. M., and Russell, P. L. 1985. Amyloglucose-catalysed erosion of native, surface-modified and chlorine-treated wheat starch granules: The influence of surface protein. *J. Cereal Sci.* **3,** 279–293.

Greenwood, C. T., and Thomson, J. 1961. Physicochemical studies on starches. Part XXIII. Some physical properties of floridean starch and the characterization of structure-type of branched α-1,4-glucans. *J. Chem. Soc.* 1534–1537.

Guan, H. P., Baba, T., and Preiss, J. 1994a. Expression of branching enzyme I of maize endosperm in *Escherichia coli. Plant Physiol.* **104,** 1449–1453.

Guan, H. P., Baba, T., and Preiss, J. 1994b. Expression of branching enzyme II of maize endosperm in *Escherichia coli. Cellul. and Molec. Biol.* **40,** 981–987.

Guan H. P., Kuriki, T., Sivak, M., and Preiss, J. 1995. Maize branching enzyme catalyzes synthesis of glycogen-like polysaccharide in glgB-deficient *Escherichia coli. Proc. Natl. Acad. Science USA* **92,** 964–967.

Guan, H. P., and Preiss J. 1993. Differentiation of the properties of the branching isozymes from maize (*Zea mays*). *Plant Physiol.* **102,** 1269–1273.

Guthrie, R. D., and Honeyman, J. 1968. "Introduction to the Chemistry of Carbohydrates." 3rd ed. Clarendon Press, Oxford, England.

Hall, D. O. 1972. Nomenclature for isolated chloroplasts. *Nature New Biol.* **235,** 125–126.

Hallberg, M., and Larsson, C. 1983. Metabolism of labelled 3-phosphoglycerate with mesophyll protoplasts and purified mesophyll chloroplasts from the C4 plant *Digitaria sanguinalis.* In "*Proceedings of the VIth International Congress on Photosynthesis*" (C. Sybesma, ed.), Vol. III., pp. 417–420. The Hague, M. Nijhoff/W. Junk.

Hammond, J. B., and Preiss, J. 1983. Spinach leaf intra and extrachloroplastic phosphorylase activities during growth. *Plant Physiol.* **73,** 709–712.

Hanes, C. S. 1940. The reversible formation of starch from glucose-1-phosphate catalyzed by potato phosphorylase. *Proc. R. Soc. Lond. Ser. B.* **120,** 174–208.

Hatch, M. D., and Osmond, C. B. 1976. Compartmentalization and transport in C_4 photosynthesis. *In* "Encyclopedia of Plant Physiology" (C. R. Stocking and U. Heber, eds.), Vol. 3, pp. 144–184. Transport in Plants III. Springer-Verlag, New York.

Hatzfeldt, W. D., and Stitt, M. 1990. A study on the rate of recycling of triose phosphates in heterotrophic *Chenopodium rubrum* cells, potato tubers, and maize endosperm. *Planta* **180,** 198–204.

Haugen, T., Ishaque, A., and Preiss, J. 1976. ADP-glucose pyrophosphorylase: Evidence for a lysine residue at the activator site of the *Escherichia coli* B enzyme. *Biochem. Biophys. Res. Commun.* **69,** 346–353.

Hawker, J. S., and Downtown, W. J. S. 1974. Starch synthetases from *Vitis vinifera* and *Zea mays. Phytochem.* **13,** 893–900.

Hawker, J. S., and Jenner, C. F. 1993. High temperature affects the activity of enzymes in the committed pathway of starch synthesis in developing wheat endosperm. *Austr. J. Plant Physiol.* **20,** 197–209.

Hawker, J. S., Ozbun, J. L., Ozaki, H., Greenberg, E., and Preiss, J. 1974. Interaction of spinach leaf adenosine diphosphate glucose α-1,4-glucan α-4-glucosyl transferase and α-1,4-glucan, α-1,4-glucan-6-glucosyl transferase in synthesis of branched α-glucan. *Arch. Biochem. Biophys.* **160,** 530–551.

Heber, U., and Heldt, H. W. 1981. The chloroplast envelope: Structure function and role in leaf metabolism. *Ann. Rev. Plant Physiol.* **25,** 393–421.

Heber, U., and Santarius, K. A. 1965. Compartmentation and reduction of pyridine nucleotides in relation to photosynthesis. *Biochim. Biophys. Acta* **109,** 390–408.

Hedman, K. D., and Boyer, C. D. 1982. Gene dosage at the *amylose-extender* locus of maize: Effects on the levels of starch branching enzymes. *Biochem. Genet.* **20,** 483–492.

Heinrich, R., and Rapoport, T. A. 1974. A linear steady-state treatment of enzymatic chains, general properties, control and effector strength. *Eur. J. Biochem.* **42,** 89–95.

Heldt, H. W. 1969. Adenine nucleotide translocation in spinach chloroplasts. *FEBS Lett.* **5,** 11–14.

Heldt, H. W. Flügge, I., and Borchert, S. 1991. Diversity of specificity and function of phosphate translocators in various plastids. *Plant Physiol.* **95,** 341–343.

Hendry, G. 1987. The ecological significance of fructan in a contemporary flora. *New Phytol.* **106,** 201–216.

Hendry, G. 1993. Evolutionary origins and natural fractions of fructans: A climatological, biogeographic and mechanistic approach. *New Phytol.* **123,** 3–14.

Hendry, G., and Wallace, R. K. 1993. The origin, distribution and evolutionary significance of fructans. *In* "Science and Technology of Fructans," (M. Susuki and N. J. Chatterton, eds.) pp. 120–139. CRC Press, Boca Raton, FL.

Henrissat, B., and Bairoch, A. 1993. New families in the classification of glycose hydrolases based on amino acid sequence similarities. *Biochem. J.* **293,** 781–788.

Hesse, H., Willmitzer, L., and Sonnewald, U. 1995. Cloning and expression analysis of sucrose-P synthase from sugar beet (*Beta vulgaris* L.) *Mol. Gen. Genet.* **247,** 515–520.

Hill, L. M., and Smith, A. M. 1991. Evidence that glucose-6-phosphate is imported as the substrate for starch synthesis by the plastids of developing pea embryos. *Planta* **185,** 91–96.

Hill, M. A., Kaufmann, K., Otero, J., and Preiss, J. 1991. Biosynthesis of bacterial glycogen: Mutagenesis of a catalytic site residue of ADPglucose pyrophosphorylase from *Escherichia coli. J. Biol. Chem.* **266,** 12455–12460.

Hixon, R. M., and Brimhall, B. 1953. The waxy cereals and starches which stain red with iodine. *In* "Starch and Its Derivatives," (J. A. Radley, ed.), pp. 252–290. Chapman and Hall, London.

Hizukuri, S. 1985. Relationship between the distribution of the chain length of amylopectin and crystalline structure of starch granule. *Carbohyd. Res.* **141,** 295.

Hizukuri, S. 1986. Polymodal distribution of the chain lengths of amylopectins and its significance. *Carbohydr. Res.* **147,** 342–347.

Hizukuri, S. 1995. Starch: Analytical aspects. *In* "Carbohydrates in Food" (A. C. Eliason, ed.), pp. 347–429. Marcel Dekker, New York.

Hizukuri, S., Fujii, M., and Nikuni, Z. 1961. Effect of temperature during germination on the crystalline type of starch in soybean seedlings. *Nature* (*Lond.*) **192,** 239–240.

Hizukuri, S., Kaneko, T., and Takeda, Y. 1983. Measurement of the chain length of amylopectin and its relevance to the origin of polymorphism of starch granules. *Biochim. Biophys. Acta.* **760,** 188–191.

Hobson, P. N., Whelan, W. J., and Peat, S. 1950. The enzymic synthesis and degradation of starch. X. The phosphorylase and Q enzyme of broad bean. The Q enzyme of wrinkled pea. *J. Chem. Soc.,* pp. 3566–3573.

Holmes, E., and Preiss, J. 1979. Characterization of *Escherichia coli* B glycogen synthase enzymatic reactions and products. *Arch. Biochem. Biophys.* **196,** 436–448.

Hopfer, U. 1997. Digestion and absorption of basic nutritional constituents. *In Textbook of Biochemistry with Clinical implications.* T. Devlin (ed.) pp. 1055–1086.

Hoseney, R. C. 1990. "Principles of Cereal Science and Technology." American Association of Cereal Chemists, St. Paul, MN.

Hovenkamp-Hermelink, J. H. M., Jacobsen, E., Ponstein, A. S., Visser, R. G. F., Vos-Scheperkeuter, G. H., Bijmolt, E. W., de Vries, J. N., Witholt, B., and Feenstra, W. J. 1987. Isolation of an amylose-free starch mutant of the potato (*Solanum tuberosum* L.). *Theor. Appl. Genet.* **75,** 217–221.

Huber, W., de Fekete, A., and Ziegler, H. 1969. Enzymes of starch metabolism in chloroplasts of the bundle sheath and palisade cells of Zea mays. *Planta* **87,** 360–364.

Hylton, C., and Smith, A. M. 1992. The rb mutation of peas causes structural and regulatory changes in ADPGlc, pyrophosphorylase from developing embryos. *Plant Physiol.* **99,** 1626–1634.

Iglesias, A. A., Barry, G. F., Meyer, C., Bloksberg, L., Nakata, P. A., Greene, T., Laughlin, M. J., Okita, T. W., Kishore, G. M., and Preiss, J. 1993. Expression of the potato tuber ADP-glucose pyrophosphorylase in *Escherichia coli. J. Biol Chem.* **268,** 1081–1086.

Iglesias, A. A., Kakefuda, G., and Preiss, J. 1991. Regulatory and structural properties of the cyanobacterial. ADPglucose pyrophosphorylases. *Plant Physiol.* **97,** 1187–1195.

Imberty, A., Buléon, A., Tran, V., and Pérez, S. 1991. Recent advances in knowledge of starch structure. *Starch/Stärke* **43,** 375–384.

Imberty, A., Chanzy, H., Pérez, S., Buléon, A., and Tran, V. 1987. New tridimensional structure for A-type starch. *Macromolecules* **20**, 2634–2636.

Imberty, A., Chanzy, H., Pérez, S., Buléon, A., and Tran, V. 1988. The double-helical nature of the crystalline part of A-starch. *J. Mol. Biol.* **20,** 2634–2636.

James, D. W., Jr., Preiss, J., and Elbein, A. D. 1985. Biosynthesis of Polysaccharides. *In* "The Polysaccharides" (G. O. Aspinall, ed.), Vol. 3, pp. 107–207. Academic Press, New York.

James, M. G., Robertson, D. S., and Meyers, A. M. 1995. Characterization of the maize gene *sugary* 1, a determinant of starch composition in kernels. *Plant Cell* **7,** 417–429.

Jefford, T. G., and Edelman, J. 1961. Changes in content and composition of the fructose polymers of *Helianthus tuberosus* L. *J. Exp. Bot.* **12,** 177–187.

Jenkins, P. J., Cameron, R. E., and Donald, A. M. 1993. A universal feature in the starch granules from different botanical sources. *Stärke* **45,** 417–420.

Jenner, C. F. 1982. Storage of starch. *In* "Encyclopedia of Plant Physiology Vol. 13A. Intracellular Carbohydrates" (F. A. Loewus and W. Tanner, eds.) pp. 700–747. Springer-Verlag, Berlin, Germany.

Jenner, C. F. 1994. Starch synthesis in the kernel of wheat under high termperature conditions. *Aust. J. Plant Physiol.* **21,** 791–806.

Jespersen, H. M., MacGregor, A., Henrissat, B., Sierks, M. R., and Svensson, B. 1993. Starch- and glycogen-debranching and branching enzymes; prediction of structural features of the catalytic $(\beta/\alpha)_8$–barrel domain and evolutionary relationship to other amylolytic enzymes. *J. Protein Chem.* **12,** 781–805.

Journet, E. P., Bligny, R., and Douce, R. 1986. Biochemical changes during sucrose deprivation in higher plant cells. *J. Biol. Chem.* **261,** 3193–3199.

Journet, E. P., and Douce, R. 1985. Enzymatic capacities of purified cauliflower bud plastids for lipid synthesis and carbohydrate metabolism. *Plant Physiol.* **79,** 458–467.

Kacser, H. 1987. Control of metabolism. *In* "The Biochemistry of Plants" (D. D. Davies, ed.), Vol. 11, pp. 39–67. Academic Press, New York.

Kacser, H., and Burns, J. A. 1973. Control of flux. *Symp. Soc. Exper. Biol.* **27,** 65–107.

Kainuma, K. 1984. Starch oligosaccharides: linear, branched and cyclic. *In* "Starch Chemistry and Technology" (R. L. Whistler, J. N. BeMiller, and E. F. Paschall, eds.), 2nd ed. pp. 125–152. Academic Press, New York.

Kainuma, K. 1988. Structure and chemistry of the starch granule. *In* "The Biochemistry of Plants" (J. Preiss, ed.), Vol. 14, pp. 141–180. Academic Press, New York.

Kainuma, K., and French, D. 1972. Naegeli amylodextrin and its relationships to starch granule structure. II. Role of water in crystallization of B-starch. *Biopolymers* **11,** 2241.

Kakefuda, G., Charng, Y.-Y., Iglesias, A. A., McIntosh, L., and Preiss, J. 1992. Molecular cloning and sequencing of ADP-glucose pyrophosphorylase from *Synechocystis* PCC6803. *Plant Physiol.* **99,** 344–347.

Kakefuda, G., and Preiss, J. 1997. Partial purification and characterization of a diurnally fluctuating novel endoamylase from *Arabidopsis thaliana* leaves. *Plant Physiol. Biochem.* **35,** 907–913.

Karrer, E. E., and Rodriguez, R. L. 1992. Metabolic regulation of rice α-amylase and sucrose synthase genes in planta. *Plant J.* **2,** 517–523.

Kasemsuwan, T., and Jane, J. 1994. Location of amylose in normal starch granules. II. Location of phosphodiester cross-linking revealed by phosphorus-31 nuclear magnetic resonance. *Cereal Chem.* **71,** 282–286.

Katz, F. R. 1991. *In* "Biotechnology and Food Ingredients" (I. Goldberg and R. Williams, eds.), pp. 315–326. Van Nostrand Reinhold, New York.

Keeling, P. L., Bacon, P. J., and Holt, D. C. 1993. Elevated temperature reduces starch deposition in wheat endosperm by reducing the activity of soluble starch synthase. *Planta* **191,** 342–348.

Keeling, P. L., Banisadr, R., Barone, L., Wasserman, B. P., and Singletary, G. W. 1994. Effects of temperature on enzymes in the pathway of starch biosynthesis in developing wheat and maize grain. *Aust. J. Plant Physiol.* **21,** 807–827.

Keeling, P. L., Wood, J. R., Tyson, R. H., and Bridges, I. G. 1988. Starch biosynthesis in developing wheat grain: Evidence against the direct involvement of triose phosphates in the metabolic pathway. *Plant Physiol.* **87,** 311–319.

Keiller, D. R., Paul, M. J., and Cockburn, W. 1987. Regulation of reserve carbohydrate metabolism in *Mesembryanthemum crystallinum* exhibiting C3 and CAM photosynthesis. *New Phytol.* **107,** 1–13.

Khoshnoodi, J., Ek, B., Rask, L., and Larsson, H. 1993. Characterization of the 97 and 103-KDa forms of starch branching enzymes from potato tubers. *FEBS Lett.* **332,** 132–138.

Kiel, J. A. K. W., Boels, J. M., Beldman, G., and Venema, G. 1990. Nucleotide sequence of the *Synechococcus* sp. PCC7942 branching enzyme (*glg*B) expression in *Bacillus subtilis. Gene.* **89,** 77–84.

Kiel, J. A. K. W., Elgersma, H. S. A., Beldman, G., Vossen, J. P. M. J., and Venema, G. 1989. Cloning and expression of the branching enzyme gene (*glg*B) from the *Cyanobacterium Synechococcus* sp. PCC7942 in *Escherichia coli. Gene* **78,** 9–17.

Kiesselbach, T. A. 1949. The structure and reproduction of corn. *Nebr. Agric. Exp. Stn. Res. Bull.* **161,** 1–96.

Kim, W. T., Franceschi, V. R., Okita, T. W., Robinson, N. L., Morell, M., and Preiss, J. 1989. Immunocytochemical localization of ADPglucose pyrophosphorylase in developing potato tuber cells. *Plant Physiol.* **91,** 217–220.

Kiss, J. Z., Hertel, R., and Sack, F. D. 1989. Amyloplasts are necessary for full gravitropic sensitivity in roots of *Arabidopsis thaliana*. *Planta.* **177,** 198–206.

Kleczkowski, L. A., Villand, P., Lüthi, E., Olsen, O.-A., and Preiss, J. 1993. Insensitivity of barley endosperm ADPGlc PPase pyrophosphorylase to 3-phosphoglycerate and orthophosphate regulation. *Plant Physiol.* **101,** 179–186.

Klösgen, R. F., Gierl, A., Schwarz-Sommer, Z., and Saedler, H. 1986. Molecular analysis of the *waxy* locus of *Zea mays*. *Mol. Gen. Genet.* **203,** 237–244.

Kossmann, J., Visser, R. F., Müller-Röber, B. T., Willmitzer, L., and Sonnewald, U. 1991. Cloning and expression analysis of a potato cDNA that encodes branching enzyme: Evidence for coexpression of starch biosynthetic genes. *Mol. Gen. Genet.* **230,** 39–44.

Koch, K. E. 1996. Carbohydrate-modulated gene expression in plants. *Annu. Rev. Plant Physiol. Plant Mol. Biol.* **47,** 509–540.

Koch, K. E., Nolte, K. D., Duke, E. R., McCarty, D. R., and Avigne, W. T. 1992. Sugar levels modulate differential expression of maize sucrose synthase genes. *Plant Cell* **4,** 59–69.

Koch, K. E., Xu, J., Duke, E. R., McCarty, D. R., and Yuan, C. X. 1995. Sucrose provides a long distance signal for coarse control of its genes affecting its metabolism. *In* "Sucrose Metabolism, Biochemistry and Molecular Biology" (H. G. Pontis, G. Salerno, and E. Eccheverria, eds.), pp. 266–277. American Society of Plant Physiologists, Rockville, MD.

Koizumi, K., Kubota, Y., Tanimoto, T., and Okada, Y. 1991. Estimation of the distribution of chain length of amylopectin by high-performance liquid chromatography with pulse amperometric detection. *J. Chromatogr.* **585,** 233.

Kram, A. 1995. "Structure and Biosynthesis of Starch: The Localization of Starch Synthesizing Enzymes with the Use of Immunoelectron Microscopy." PhD thesis. University of Groningen, The Netherlands.

Kram, A., Oostergetel, G. T., and van Bruggen, E. F. J. 1993. Localization of branching enzyme in potato tuber cells with the use of immunoelectron microscopy. *Plant Physiol.* **101,** 237–243.

Krapp, A., and Stitt, M. 1995. An evaluation of direct and indirect mechanisms for the "Sink regulation" of photosynthesis of spinach: Changes in gas exchange, carbohydrates metabolites, enzyme activities and steady state transcript levels after cold-girdling source leaves. *Planta* **195,** 313–323.

Krishnan, H. B., Reeves, C. D., and Okita, T. W. 1986. ADPglucose pyrophosphorylase is encoded by different mRNA Transcripts in leaf and endosperm of cereals. *Plant Physiol.* **81,** 642–645.

Krisman, C. R. 1962. A method for the colorimetric estimation of glycogen with iodine. *Anal. Biochem.* **4,** 17–23.

Krisman, C. R. 1972. A possible intermediate in the initiation of glycogen biosynthesis. *Biochem. Biophys. Res. Commun.* **46,** 1206–1216.

Kruckeberg, A. L., Neuhaus, H. E., Feil, R., Gottlieb, L. D., and Stitt, M. 1989. Decreased-activity mutants of phosphoglucose isomerase in the cytosol and chloroplast of Clarkia xantiana: Impact on mass-action ratios and fluxes to sucrose and starch and estimation of flux control coefficients and elasticity coefficients. *Biochem. J.* **261,** 457–467.

Kruger, N. J., and ap Rees, T. 1983. Properties of α-glucan phosphorylase from pea chloroplasts. *Biochemistry* **22,** 1891–1898.

Kumar, A., Larsen, C. E., and Preiss, J. 1986. Biosynthesis of bacterial glycogen. Primary structure of *Escherichia coli* ADPglucose: α-1,4-glucan, 4-glycosyltransferase deduced from the nucleotide sequence of the glg A gene. *J. Biol. Chem.* **261,** 16256–16259.

Kumar, A., Ghosh, P., Lee, Y. M., Hill, M. A., and Preiss, J. 1989. Biosynthesis of bacterial glycogen: determination of the amino acid changes that alter the regulatory properties of a mutant *Escherichia coli* ADPglucose synthetase. *J. Biol. Chem.* **264,** 10464–10471.

Kumar, A., Tanaka, T., Lee, Y. M., and Preiss, J. 1988. Biosynthesis of bacterial glycogen: Use of site-directed mutagenesis to probe the role of tyrosine 114 in the catalytic mechanism of ADPglucose synthetase from *Escherichia coli. J. Biol. Chem.* **263,** 14634–14639.

Kuriki, T., Guan, H., Sivak, M. N., and Preiss, J. 1996. Analysis of the active center of branching enzyme II from maize endosperm. *J. Prot. Chem.* **15,** 305–313.

Kuriki, T., Stewart, D. C., and Preiss, J. 1997. Construction of chimeric enzymes out of maize endosperm branching enzymes I and II: Activity and properties. *J. Biol. Chem.* **272,** 28999–29004.

Lavintman, N, and Cardini, C. E. 1973. Particulate UDP-glucose : protein transglucosylase from potato tuber. *FEBS Lett.* **29,** 43–46.

Lavintman, N., Tandecarz, J., Carceller, M., Mendiara, S., and Cardini, C. E. 1974. Role of uridine diphosphate glucose in the biosynthesis of starch: Mechanism of formation and enlargement of a glucoproteic acceptor. *Eur. J. Biochem.* **50,** 145–155.

Lea, P. J., and Leegood, R. C. 1993. *In* "Plant Biochemistry and Molecular Biology." John Wiley & Sons.

Lee, Y. M., Marshall, and Whelan, 1971. The substrate specificity of amylopectin-debranching enzymes from sweet corn. *Arch. Biochem. Biophys.* **143,** 365.

Lee, Y. M., and Preiss, J. 1986. Covalent modification of substrate binding sites of *E. coli* ADPglucose synthetase: Isolation and structural characterization of 8-azido ADPglucose incorporated peptides. *J. Biol. Chem.* **261,** 1058–1064.

Leegood, R. C. 1985. The inter-cellular compartmentation of metabolites in leaves of *Zea mays. Planta.* **164,** 163–171.

Leegood, R. C. 1996. Primary photosynthetic production: Physiology and metabolism. *In* "Photoassimilate distribution in Plants and Crops. Source–Sink Relationships" (E. Zamsky, and A. A. Schaeffer, eds.), pp. 21–42. Marcel Dekker, New York.

Leegood, R. C., and von Caemmerer, S. 1988. The relationship between content of photosynthetic metabolites and the rate of photosynthetic carbon assimilation in leaves of *Amaranthus edulis. Planta.* **174,** 253–262.

Leloir, L. F., and Cardini, C. E. 1955. The biosynthesis of sucrose phosphate. *J. Biol. Chem.* **214**, 157–165.

Leloir, L. F., de Fekete, M. A. R., and Cardini, C. E. 1961. Starch and oligosaccharide synthesis from uridine diphosphate glucose. *J. Biol. Chem.* **236,** 636–641.

Lemesle-Varloot, L., Henrissat, B., Gaboriaud, C., Bissery, V., Morgat, A., and Mornon, J. P. 1990. Hydrophobic cluster analysis: procedures to derive structural and functional information from 2-D-representation of protein sequences. *Biochimie* **72,** 555–574.

Leung, P., Lee, Y. M., Greenberg, E., Esch, K., Boylan, S., and Preiss, J. 1986. Cloning and expression of the *Escherichia coli glg* C gene from a mutant containing an ADPglucose pyrophosphorylase with altered allosteric properties. *J. Bacteriol.* **167,** 82–88.

Leung, P., and Preiss, J. 1987. Biosynthesis of bacterial glycogen. Primary structure of *Salmonella typhimurium* ADPglucose synthetase as deduced from the nucleotide sequence of the *glg* C gene. *J. Bacteriol.* **169,** 4355–4360.

Levi, C., and Preiss, J. 1976. The regulatory properties of the ADPglucose pyrophosphorylase of the blue-green bacterium, *Synechococcus* 6301. *Plant Physiol.* **58,** 753–756.

Levi, C., and Preiss, J. 1978. Amylopectin degradation in pea chloroplast extracts. *Plant Physiol.* **61,** 218–220.

Li, L., and Preiss, J. 1992. Characterization of ADPglucose pyrophosphorylase from a starch-deficient mutant of *Arabidopsis thaliana* (L.). *Carbohyd. Res.* **227,** 227–239.

Libessart, N., Maddelein, M. L., Van den Koornhuise, N., Decq., A., Delrue, B., and Ball, S. 1995. Storage, photosynthesis and growth: The conditional nature of mutations affecting starch synthesis and structure in *Chalamydomonas reinhardtii. Plant Cell* **7,** 1117–1127.

Lim, S., and Seib, P. A. 1993. Location of phosphate esters in wheat starch phosphate by 31P-nuclear magnetic resonance spectroscopy. *Cereal Chem.* **70,** 145.

Lin, T. P., Caspar, T., Somerville, C., and Preiss, J. 1988a. A starch deficient mutant of *Arabidopsis thaliana* with low ADPglucose pyrophosphorylase activity lacks one of the two subunits of the enzyme. *Plant Physiol.* **88,** 1175–1181.

Lin, T. P., Caspar, T., Somerville, C., and Preiss, J. 1988b. Isolation and characterization of a starchless mutant of *Arabidopsis thaliana* L. Henyh lacking ADPglucose pyrophosphorylase activity. *Plant Physiol.* **86,** 1131–1135.

Loewus, F., and Labarca, C. 1973. *In* "Biogenesis of the Plant Cell Wall Polysaccharides" (F. Loewus, ed.) pp. 175–193. Academic Press, New York.

Lomako, J., Lomako, W. M., and Whelan, W. J. 1988. A self-glucosylating protein is the primer for rabbit muscle glycogen biosynthesis. *FASEB. J.* **2,** 3097–3103.

Love, J., Mackie, W., McKinell, J. W. and Percival, E. 1963. *J. Chem. Soc.* 4177–4182.

Macdonald, F. D., and ap Rees, T. 1983. Enzymatic properties of amyloplasts from suspension cultures of soybean. *Biochim. Biophys. Acta.* **755,** 81–89.

Macdonald, F. D., and Preiss, J. 1983. Solubilization of the starch-granule-bound starch synthases of normal maize kernels. *Plant Physiol.* **73,** 175–178.

Macdonald, F. D., and Preiss, J. 1985. Partial purification and characterization of granule-bound starch synthases from normal and *waxy* maize. *Plant. Physiol.* **78,** 849–852.

Macheral, D., Kobayashi, H., Akazawa, T., Kawano, S., and Kuioiwa, T. 1985. Amyloplast nucleoids in sycamore cells and presence of amyloplast DNA of homologous sequences to chloroplast genes. *Biochem. Biophys. Res. Commun.* **133,** 140–146.

MacMasters, M. M. 1964. Microscopic techniques for determining starch granule properties. *In* "Methods in Carbohydrate Chemistry." (R. L. Whistler, ed.), Academic Press, New York.

Maddelein, M.-L., Bellanger, F., Delrue, B., Libessart, N., D'hulst, C., Van den Koornhuyse, N., Fontaine, T., Wieruszeski, J.-M., Decq. A., and Ball, S. 1994a. Genetic dissection of starch metabolism in the monocellular alga *Chlamydomonas reinhardtii:* Determination of granule-bound and soluble starch synthases functions in amylopectin biosynthesis. *J. Biol. Chem.* **269,** 25150–25157.

Maddelein, M.-L., Libessart, N., Bellanger, F. Delrue, B., D'Hulst, C., Van Den Koornhuyse, N., Fontaine, T., Wieruszeski, J.-M., T. M., Decq, A., and Ball, S. 1994b. Toward an understanding of the biogenesis of the starch granule: Determination of granule-bound and soluble starch synthase functions in amylopectin synthesis. *J. Biol. Chem.* **269,** 25150–255157.

Manners, D. J. 1985. Starch. *In* "Biochemistry of Storage Carbohydrates in Green Plants" (P. M. Dey and K. A. Dixon, eds.), pp. 149–203. Academic Press, New York.

Manners, D. J. 1989. Recent developments in our understanding of amylopectin structure. *Carbohyd. Polym* **11,** 87–112.

Manners, D. J. 1997. Observations on the specificity and nomenclature of starch debranching enzymes. *J. Appl. Glycosci.* **44,** 83–85.

Manners, D. J., and Matheson, N. K. 1981. The fine structure of amylopectin. *Carbohydr. Res.* **90,** 99–110.

Manners, D. J., and Sturgeon, R. J. 1982. Reserve carbohydrates of algae, fungi and lichens. *In* "Encyclopedia of Plant Physiology: Vol. 13A. Intracellular Carbohydrates" (F. A. Loewus and W. Tanner, eds.) pp. 472–514. Springer-Verlag, Berlin, Germany.

Marcus, A. 1989. "Molecular Biology. Vol. 15. The Biochemistry of Plants, a Comprehensive Treatise." Academic Press, New York.

Mares, D. J., Hawker, J. S., and Possingham, J. V. 1978. Starch synthesizing enzymes in chloroplasts of developing leaves of spinach (*Spinacea oleracea* L.). *J. Exp. Bot.* **29,** 829–835.

Mbaku, S. B., Fritz, G. J., and Bowes, G. 1978. Photosynthetics and carbohydrate metabolism in isolated leaf cells of *Digitaria pentzii. Plant Physiol.* **62,** 510–515.

Meeuse, B. J. D., Andries, M., and Wood, J. A. 1960. *J. Exp. Bot.* **11,** 129–140.

Meeuse, J. D., and Kreger, D. R. 1954. On the nature of floridean starch and *Ulva* starch. *Biochim. Biophys. Acta.* **13,** 593–595.

Meier, H., and Reid, J. S. G. 1982. Reserve polysaccharides other than starch in higher plants. *In* "Encyclopedia of Plant Physiology: Vol. 13A. Intracellular Carbohydrates" (F. A. Loewus and W. Tanner, eds.), pp. 418–471. Springer-Verlag, Berlin, Germany.

Meuser, F., Manners, D. J. and Seibel, W. 1993. "Plant Polymeric Carbohydrates" CRC Press, Boca Raton, FL.

Mengel, K., and Judel, G.-K. 1981. Effect of light intensity on the activity of starch synthesizing enzymes and starch synthesis in developing wheat grains. *Physiol. Plant* **51,** 13–18.

Miller, M. E., and Chourey, P. 1995. Intracellular immunolocalization of adenosine 5′-diphosphoglucose pyrophosphoylase in developing endosperm cells of maize (*Zea mays* L.). *Planta* **197,** 522–527.

Mizuno, K., Kawasaki, T., Shimada, H., Satoh, H., Kobayashi, E., Okamura, S., Arai, Y., and Baba, T. 1993. Alteration of the structural properties of starch components by the lack of an isoform of starch branching enzyme in rice seeds. *J. Biol. Chem.* **268,** 19084–19091.

Mizuno, K., Kimura, K., Arai, Y., Kawasaki, T., Shimada, H., and Baba, T. 1992. Starch branching enzymes from immature rice seeds. *J. Biochem.* **112,** 643–651.

Mohabir, G., and John, F. 1988. Effect of temperature on starch synthesis in potato tissue and in amyloplasts. *Plant Physiol.* **88,** 1222–1228.

Moore, C. J., and Turner, J. F. 1969. ADP-glucose pyrophosphorylase in wheat grains. *Nature (London)* **223,** 303–304.

Morell, M. K., Bloom, M., Knowles, V., and Preiss, J. 1987. Subunit structure of spinach leaf ADPglucose pyrophosphorylase. *Plant Physiol.* **85,** 185–187.

Morell, M., Bloom, M., Larsen, R., Okita, T. W., and Preiss, J. 1987. Biochemistry and molecular biology of starch synthesis. *In* "Plant Gene Systems and Their Biology" (J. L. Key and L. McIntosh, eds.), pp. 227–242. Alan R. Liss, New York.

Morell, M., Bloom, M., and Preiss, J. 1988. Affinity labeling of the allosteric activator site(s) of spinach leaf ADP-glucose pyrophosphorylase. *J. Biol. Chem.* **263,** 633–637.

Moreno, S., Cardini, C. E., and Tandecarz, J. S. 1986. α-Glucan synthesis on a protein primer. A reconstituted system for the formation of protein-bound α-glucan. *Eur. J. Biochem.* **162,** 609–614.

Morrison, W. R., and Karkalas, J. 1990. Starch. *In* "Methods in Plant Biochemistry," (P. M. Dey, ed.), pp. 323–352. Vol. 2, Academic Press, New York.

Mouille, G., Maddelein, M.-L., Libessart, N., Talaga, P., Decq, A., Delrus, B., and Ball, S. 1996. Preamylopectin processing: A mandatory step for starch biosynthesis in plants. *Plant Cell* **8,** 1353–1366.

Mount, S. M. 1982. A catalogue of splice junction sequences. *Nucl. Acid Res.* **10,** 459–462.

Mu, C., Harn, C., Ko, Y.-T., Singletary, G. W., Keeling, P. L., and Wasserman, B. P. 1994. Association of a 76 KDa polypeptide with soluble starch synthase I activity in maize (CV73) endosperm. *Plant J.* **6,** 151–159.

Mu-Forster, C., Huang, R., Powers, J. R., Harriman, R. W. Knight, M., Singletary, G. W., Keeling, P., and Wasserman, B. P. 1996. Physical association of starch biosynthetic enzymes with starch granules of maize endosperm. *Plant Physiol.* **111,** 821–829.

Mulichak, A. M., Skrzypczak-Jankum, E., Rydel, T. J., Tulinsky, A, and Preiss, J. 1988. Crystallization and preliminary diffraction data, of *Escherichia coli* ADPglucose pyrophosphorylase. *J. Biol. Chem.* **263,** 17237–17238.

Müller-Röber, B. and Kossman, J. 1994. Approaches to influence starch quantity and starch quality in transgenic plants. *Plant Cell Environ.* **17,** 601–613.

Müller-Röber, B. T., Kossman, J., Hannah, L. C., Willmitzer, L., and Sonnewald, U. 1990. One of two different ADP-glucose pyrophosphorylase genes from potato responds strongly to elevated levels of sucrose. *Mol. Gen. Genet.* **224,** 136–146.

Müller-Röber, B. T., Sonnewald, U., and Willmitzer, L. 1992. Inhibition of ADPglucose pyrophosphorylase in transgenic potatoes leads to sugar-storing tubers and influences tuber formation and expression of tuber storage protein genes. *EMBO J.* **11,** 1229–1238.

Nakamura, Y., Umemoto, T., Ogata, N., Kuboki, Y., Yano, M., and Saski, T. 1996a. Starch debranching enzyme (R-enzyme or pullulanase) from developing rice endosperm: Purification, cDNA and chromosomal localization of the gene. *Planta* **199,** 203–218.

Nakamura, Y., Umemoto, T., Takahata, Y., Komae, K., Amano, E., and Satoh, H. 1996b. Changes in structure of starch and enzyme activities affected by *sugary* mutation in developing rice endosperm: Possible role of starch debranching enzyme in amylopectin biosynthesis. *Physiol. Plant.*

Nakamura, Y., and Yamanouchi, H. 1992. Nucleotide sequence of a cDNA encoding starch-branching enzyme or Q-enzyme I from rice endosperm. *Plant Physiol.* **99,** 1265–1266.

Nakata, P. A., Greene, T. W., Anderson, J. M., Smith-White, B. J., Okita, T. W., and Preiss, J. 1991. Comparison of the primary sequences of two potato tuber ADP-glucose pyrophosphorylase subunits. *Plant Mol. Biol.* **17,** 1089–1093.

Nelson, O. 1985. Genetic control of starch synthesis in maize endosperm—a review. *In* "New Approaches to Research in Cereal Carbohydrates." (R. D. Hill and L. Munck, eds.) Elsevier Science Pub, Amsterdam, pp. 19–28.

Nelson, O. E. 1962. The *waxy* locus in maize. I. Intralocus recombination frequency estimates by pollen and by conventional analyses. *Genetics* **47,** 737–742.

Nelson, O. E. 1987. The *waxy* locus in maize twenty five years later. *Genetics* **116,** 339–342.

Nelson, O. E., Chourey, P. S., and Chang, M. T. 1978. Nucleoside diphosphate sugar-starch glucosyl transferase activity in *wx* starch granules. *Plant Physiol.* **62,** 383–386.

Nelson, O. E., and Pan, D. 1995. Starch synthesis in maize endosperms. Ann. Rev. *Plant Physiol. Mol. Biol.* **46,** 475–496.

Nelson, O. E., and Rines, H. W. 1962. The enzyme deficiency in the *waxy* mutant of maize. *Biochem. Biophys. Res. Commun.* **9,** 297–300.

Neuffer, M. G., Coe, E. H., and Wessler, S. R. 1997. "Mutants of Maize." Cold Spring Harbor Laboratory Press, Cold Spring Harbor, New York.

Neuhaus, H. E., Henrichs, G., and Scheibe, R. 1993. Characterization of glucose-6-P incorporation into starch by isolated intact cauliflower-bud plastids. *Plant Physiol* **101,** 573–578.

Neuhaus, H. E., Kruckeberg, A. L., Feil, R., and Stitt, M. 1989. Reduced activity mutants of phosphoglucose isomerase in the cytosol and chloroplast of *Clarkia xantiana.* II. Studies of the mechanisms which regulate photosynthate partitioning. *Planta* **178,** 110–122.

Neuhaus, H. E., and Stitt, M. 1990. Control analysis of photosynthate partitioning: Impact of reduced activity of ADPglucose pyrophosphorylase or plastid phosphoglucomutase on the fluxes to starch and sucrose in *Arabidopsis. Planta.* **182,** 445–454.

Nishimura, M., and Beevers, H. 1979. Subcellular distribution of gluconeogenic enzymes in germinating castor bean endosperm. *Plant Physiol.* **64,** 31–37.

Nishimura, M., Graham, D., and Akazawa, T. 1976. Isolation of intact chloroplasts and other cell organelles from spinach leaf protoplasts. *Plant Physiol.* **58,** 309–314.

Obata-Sasamoto, H., and Suzuki, H. 1979. Activities of enzymes relating to starch synthesis and endogenous levels of growth regulators in potato stolon tips during tuberization. *Physiol. Plant* **45,** 320–324.

Ohsugi, R., and Huber, S. C. 1987. Light modulation and localization of sucrose phosphate synthase activity between mesophyll cells and bundle sheath cells in C4 species. *Plant Physiol.* **84,** 1096–1101.

Okamoto, K., and Akazawa, T. 1979. Enzymatic mechanism of starch breakdown in germination of rice *Oryza sativa* cultivar kimmaze seeds. 8. Immunohistochemical localization of β-amylase. *Plant Physiol.* **64,** 337–340.

Okita, T. 1992. Is there an alternative pathway for starch synthesis? *Plant Physiol.* **100,** 560–564.

Okita, T. W., Greenberg, E., Kuhn, D. N., and Preiss, J. 1979. Subcellular localization of the starch degradative and biosynthetic enzymes of spinach leaves. *Plant Physiol.* **64,** 187–192.

Okita, T. W., Nakata, P. A., Anderson, J. M., Sowokinos, J., Morell, M., and Preiss, J. 1990. The subunit structure of potato tuber ADPglucose pyrophosphorylase. *Plant Physiol.* **93,** 785–790.

Okuno, K., and Sakaguchi, S. 1982. Inheritance of starch characteristics in perisperm of *Amaranthus hypochondriacus. J. Hered.* **73,** 467.

Olive, M. R., Ellis, R. J., and Schuch, W. W. 1989. Isolation and nucleotide sequence of cDNA clones encoding ADPglucose pyrophosphorylase polypeptides from wheat leaf and endosperm. *Plant Mol. Biol.* **12,** 525–538.

Oostergetel, G. T., and van Bruggen, E. F. J. 1989. On the origin of a low angle spacing in starch. *Staërke* **41,** 331–335.

Oostergetel, G. T., and van Bruggen, E. F. J. 1993. The crystalline domains in potato starch granules are arranged in a helical fashion. *Carbohyd. Polym.* **21,** 7–12.

Ou-Lee, T.-M., and Setter, T. 1985a. Effect of increased temperature in apical regions of maize ears on starch synthesis enzymes and accumulation of sugars and starch. *Plant Physiol.* **79,** 852–855.

Ou-Lee, T.-M., and Setter, T. 1985b. Enzyme activities of starch and sucrose pathways and growth of apical and basal maize kernels. *Plant Physiol.* **79,** 848–851.

Ozbun, J. L., Hawker, J. S., Greenberg, E., Lammel, C., Preiss, J., and Lee, E. Y. C. 1973. Starch synthetase, phosporylase, ADPglucose pyrophosphorylase and UDP-glucose pyrophosphorylase in developing maize kernels. *Plant Physiol.* **51,** 1–5.

Ozbun, J. L., Hawker, J. S., and Preiss, J. 1971. Multiple forms of α-1,4 glucan synthetase from spinach leaves. *Biochem. Biophys. Res. Comun.* **43,** 631–636.

Paladini, A. C. 1995. The early days of the Instituto de Investigaciones Bioquimicas, Fundacion Campomar. *In* "Sucrose Metabolism, Biochemistry, Physiology, and Molecular Biology" (H. Pontis, G. Salerno, and E. Echeverria, eds.), pp. 1–3. American Society of Plant Physiologists, Rockville, MD.

Pan, D., and Nelson, O. E. 1984. A debranching enzyme deficiency in endosperms of the *sugary* 1 mutants of maize. *Plant Physiol.* **74,** 324–328.

Peat, S., Whelan, W. J., and Thomas, G. J. 1952. The enzymic synthesis and degradation of starch. *J. Chem. Soc.* 4546–4548.

Percival, E., and McDowell, R. H. 1985. Algal polysaccharides. *In* "Biochemistry of Storage Carbohydrates in Green Plants" (P. M. Dey and R. A. Dixon, eds.), pp. 305–348. Academic Press, New York.

Perez, C. M., Perdon, A. A., Resurreccion, A. P., Villareal, R. M., and Juliano, B. O. 1975. Enzymes of carbohydrate metabolism in developing rice grain. *Plant Physiol.* **56,** 579–586.

Pettersson, G., and Ryde-Pettersson, U. 1989. Metabolites controlling the rate of starch synthesis in the chloroplast of C_3 Plants. *Eur. J. Biochem.* **179,** 169–172.

Pitcher, J., Smythe, C., Campbell, D. G., and Cohen, P. 1987. Identification of the 38-kDa subunit of rabbit skeletal muscle glycogen synthase as glycogenin. *Eur. J. Biochem.* **169,** 497–502.

Pitcher, J., Smythe, C., and Cohen, P. 1988. Glycogenin is the priming glucosyltransferase required for the initiation of glycogen biogenesis in rabbit skeletal muscle. *Eur. J. Biochem.* **176,** 391–395.

Plaxton, W. C., and Preiss, J. 1987. Purification and properties of nonproteolytically degraded ADPglucose pyrophosphorylase from maize endosperm. *Plant Physiol.* **83,** 105–112.

Pollock, C. J., Cairns, A. J., Sims, I. M., and Housley, T. L. 1996. *In* "Photoassimilate Distribution in Plants and Crops: Source–Sink Relationships" (E. Zamsky and A. A. Schaeffer, eds.), pp. 97–113. Marcel Dekker, New York.

Pollock, C. J., and Chatterton, N. J. 1988. Fructans. *In* "The Biochemistry of Plants: Vol. 14. Carbohydrates" (J. Preiss, ed.), pp. 109–140.

Pollock, C., and Preiss, J. 1980. The citrate-stimulated starch synthase of starchy maize kernels: Purification and properties. *Arch. Biochem. Biophys.* **204,** 578–588.

Pontis, H. G., and del Campillo, E. 1985. Fructans. *In* "The Biochemistry of Storage Carbohydrates in Green Plants" (P. M. Dey and R. A. Dixon, eds.), pp. 205–227. Academic Press, New York.

Pontis, H. G., Salerno, G. L., and Echeverria, E. J., eds. 1995. "Sucrose Metabolism, Biochemistry, Physiology and Molecular Biology" American Society of Plant Physiologists, Rockville, MD.

Pozueta-Romero, J., Frehner, M., Viale, A. M., and Akazawa, T. 1991. Comparative analysis of mitochondrial and amyloplast adenylate translocators. *FEBS Lett.* **287,** 62–66.

Preiss, J. 1969. The regulation of the biosynthesis of α-1,4 glucans in bacteria and plants. *In* "Current Topics of Cellular Regulation" (E. R. Stadtmen and B. L. Horecker, eds.) Vol. 1, pp. 125–161. Academic Press, New York.

Preiss, J. 1982a. Biosynthesis of starch and its regulation. *In* "Encyclopedia of Plant Physiology: Plant Carbohydrates" (W. Tanner and F. Loewus, eds.), Vol. 13A, pp. 397–417. Springer-Verlag, Berlin, Germany.

Preiss, J. 1982b. Regulation of the biosynthesis and degradation of starch. *Ann. Rev. Plant Physiol.* **54,** 431–454.

Preiss, J. 1984. Bacterial glycogen and its regulation. *Ann. Rev. Microbiol.* **38,** 419–458.

Preiss, J. 1988. Biosynthesis of starch and its regulation. *In* "The Biochemistry of Plants" (J. Preiss, ed.), pp. 181–254. Academic Press, New York.

Preiss, J. 1991. Starch biosynthesis and its regulation. *In* "Oxford Surveys of Plant Molecular and Cell Biology" (B. J. Miflin, ed.), Vol. 7, pp. 59–114. Oxford University Press, Oxford, England.

Preiss, J. 1996. ADPglucose pyrophosphorylase: Basic science and applications in biotechnology. *In* "Biotechnology Annual Review" (M. R. El-Gewely, ed.), pp. 259–279. Elsevier Science, New York.

Preiss, J. In press. Regulation of glycogen synthesis. *In:* "*Escherichia coli* and *Salmonella typhimurium:* Cellular and Molecular Biology" (F. C. Neidhardt, ed.), 2nd ed. Washington, DC, American Society of Microbiologist.

Preiss, J., Ball, K., Charng, Y. Y., and Iglesias, A. 1992. Structure–function relationships of ADPglucose pyrophosphorylase regulatory sites and *in vivo* evidence that ADPglucose is synthesized only in the chloroplast via ADPglucose pyrophosphorylase. *In* "Research in Photosynthesis" (N. Murata, ed.), Vol. III, pp. 697–700. Kluwer Academic Publishers, Dordrecht, The Netherlands.

Preiss, J., and Boyer, C. D. 1980. Evidence of independent control of the multiple forms of maize endosperm branching enzymes and starch synthases. *In* "Mechanisms of Saccharide Polymerization and Depolymerization" (J. J. Marshall, ed.), pp. 161–174. Academic Press, New York.

Preiss, J., Cress, D., Hutny, J., Morell, M., Bloom, M., Okita, T., and Anderson, J. 1989. Regulation of starch synthesis: Biochemical and genetic studies. *In* "Biocatalysis in Agricultural Biotechnology" (J. R. Whitaker and P. E. Sonnet, eds.), pp. 84–92. American Chemical Society.

Preiss, J., Danner, S., Summers, P. S., Morell, M., Barton, C. R., Yang, L., and Nieder, M. 1990. Molecular characterization of the brittle-2 gene effect on maize endosperm ADPglucose Pyrophosphorylase subunits. *Plant Physiol.* **92,** 881–885.

Preiss, J., Ghosh, H. P., and Wittkop, J. 1967. Regulation of the biosynthesis of starch in spinach leaf chloroplasts. *In* "Biochemistry of Chloroplasts" (T. Goodwin, ed.), Vol. II, p. 131. Academic Press, New York.

Preiss, J., and Levi, C. 1980. Starch biosynthesis and degradation. *In* "The Biochemistry of Plants" (J. Preis, ed.), Vol. 3, pp. 371–423. Academic Press, New York.

Preiss, J., Robinson, N., Spilatro, S., and McNamara, K. 1985. Starch Synthesis and Its Regulation. *In* "Regulation of Carbon Partitioning in Photosynthetic Tissue" (R. L. Heath and J. Preiss, eds.), pp. 1–26. American Society of Plant Physiologists, Rockville, MD.

Preiss, J., and Romeo, T. 1989. Physiology, biochemistry and genetics of bacterial glycogen synthesis. *Adv. Microb. Physiol.* **30,** 184–238.

Preiss, J., and Romeo, T. 1994. Molecular biology and regulatory aspects of glycogen biosynthesis in bacteria. *In* Progress nucl. acid res. molec. biology. (W. E. Cohn and K. Moldave, eds.), Vol. 47, pp. 299–329. Academic Press, New York.

Preiss, J., Sheng, J., Fu, Y., and Ballicora, M. A. 1996. *In* Proc. 10th Intl. Photosynthesis Congress (P. Mathis, ed.), Vol. 5, pp. 47–52. Kluwer Academic Publishing.

Preiss, J., and Sivak, M. N. 1996. Starch synthesis in sinks and sources. *In* "Photoassimilate Distribution in Plants and Crops: Source–Sink relationships." (E. Zamski and A. A. Schaffer, eds.), pp. 63–96. Marcel Dekker, New York.

Preiss, J., and Walsh, D. A. 1981. The comparative biochemistry of glycogen and starch. *In* "Biology of Carbohydrates" (V. Ginsburg, ed.), Vol. 1., pp. 199–314. John Wiley & Sons, New York.

Quick, W. P., and Stitt, M. 1996. Sucrose metabolism in sinks and sources. *In* "Photoassimilate Distribution in Plants and Crops: Source–Sink Relationships" (E. Zamsky and A. A. Schaeffer, eds.), pp. 115–156. Marcel Dekker, New York.

Rabinowitch, E. I. 1945. "Photosynthesis and Related Processes." Vol. 1. John Wiley & Sons (Interscience), New York.

Rabinowitch, E. I. 1956. "Photosynthesis and Related Processes." Vol. II, Part 2. John Wiley & Sons (Interscience), New York.

Recondo, E., Dankert, M., and Leloir, L. F. 1963. Isolation of adenosine diphosphate D-glucose from corn grains. *Biochem. Biophys. Res. Commun.* **12,** 204–207.

Recondo, E., and Leloir, L. F. 1961. Adenosine diphosphate glucose and starch synthesis. *Biochem. Biophys. Res. Commun.* **6,** 85–88.

Reibach, P. H., and Benedict, C. R. 1982. Biosynthesis of starch in proplastids of germinating *Ricinus communis* endosperm tissue. *Plant Physiol.* **70,** 252–256.

Reeves, C. D., Krishnan, H. B., and Okita, T. W. 1986. Gene expression in developing wheat endosperm. Accumulation of gliadin and ADPglucose pyrophosphorylase messenger RNAs and polypeptides. *Plant Physiol.* **82,** 34–40.

Rhode, W., Becker, D., and Salamini, F. 1988. Structural analysis of the waxy locus from *Hordeum vulgare. Nucl. Acids Res.* **16,** 7185–7186.

Rismeier, J. W., Flugge, U. I., Schultz, B., Heinecke, D., Heldt, H. W., Wilmitzer, L., and Frommer, W. B. 1993. Antisense repression of of the chloroplast triose phosphate translocator affects carbon partitioning in transgenic potato plants. *Proc. Natl. Acad. Sci. USA* **90,** 6160–6164.

Robin, J. P., Mercier, C., Charbonnière, R., and Guibot, A. 1974. Lintnerized starches: Gel filtration and enzymatic studies of insoluble residues from prolonged acid treatment of potato starch. *Cereal Chem.* **51,** 389–406.

Robinson, N., and Preiss, J. 1987. Localization of carbohydrate metabolizing enzymes in guard cells of *Commelina communis. Plant Physiol.* **85,** 360–364.

Robinson, S. P., and Walker, D. A. 1979. The site of sucrose synthesis in isolated leaf protoplasts. *FEBS Lett.* **107,** 295–299.

Rohde, W., Becker, D., and Salamini, F. 1988. Structural analysis of the *waxy* locus from *Hordeum vulgare. Nucl. Acids Res.* **16,** 7185–7186.

Rossman, M. G., Moras, D., and Olsen, K. W. 1974. Chemical and biological evolution of a nucleotide-binding protein. *Nature,* **250,** 194–199.

Rost, B., and Sander, R. 1993. Improved prediction of protein secondary structure by use of sequence profiles and neural networks. *Proc. Natl. Acad. Sci. USA* **90,** 7558–7562.

Sack, F. D., and Kiss, J. Z. 1989. Plastids and gravity perception. *In* "Physiology, biochemistry and genetics of nongreen plastids" (C. T. Boyer, J. C. Shannon, and R. C. Hardison, eds.), Vol 2. A.A.S.P.P. Rockville, MD.

Sanwal, G. G., Greenberg, E., Hardie, J., Cameron, E., and Preiss, J. 1968. Regulation of starch biosynthesis in plant leaves: Activation and inhibition of ADPglucose Pyrophosphorylase. *Plant Physiol.* **43,** 417–427.

Schultz, B., Frommer, W. B., Flugge, U. I., Hummel, S., Fischer, K., and Wilmitzer, L. 1993. Expression of the triose phosphate translocator gene from potato is light dependent and restricted to green tissues. *Mol. Gen. Genet.* **238,** 357–361.

Scofield, J. D., and Greenwell, P. 1987. Wheat starch granule proteins and their technological significance. *In* "Cereals in a European Context. Proc. First Conference on Food Science and Technology" (I. D. Morton, ed.), pp. 407–419. Ellis Horwood Publishers City, St.

Shannon, J. C., and Garwood, D. L. 1984. Genetics and physiology of starch development. *In* "Starch: Chemistry and Technology" (R. L. Whistler, J. N. BeMiller, and E. F. Paschall, eds.), pp. 26–86. Academic Press, New York.

Shannon, J. C., Pien, F.-M., and Lui, K.-C. 1996. *Plant Physiol.* **110,** 835–843.

Sheng, J., Charng, Y.-Y., and Preiss, J. 1996. Site-directed mutagenesis of Lysine^{382}, the activator binding site, of ADP-glucose pyrophosphorylase from *Anabaena* PCC 7120. *Biochem.* **35,** 3115–3121.

Sheng, J., and Preiss, J. 1998. Arginine^{294} is essential for the inhibition of anabaena PCC 7120 ADP-glucose pyrophosphorylase by phosphate. *Biochemistry* **36,** 13077–13084.

Sheridan, W. (ed.) 1982. Maize for biological research. *Plant Mol. Biol. Assoc.* University of North Dakota Press, 434 pp.

Shively, J. M. 1988. Inclusions: Granules of polyglucose, polyphosphate and poly-β-hydroxybutarate. *In* "The Cyanobacteria" (L. Packer, and A. N. Glazer, eds.), Vol. 167, pp. 196–199. Academic Press, New York.

Shure, M., Wessler, S., and Federoff, N. 1983. Molecular identification and isolation of the *Waxy* Locus in maize. *Cell* **35,** 225–233.

Singh, B. K., and Preiss, J. 1985. Starch branching enzymes from maize: Immunological characterization using polyclonal and monoclonal antibodies. *Plant Physiol.* **79,** 34–40.

Singh, M. B., Hough, T., Theerakulpisut, P., Avjioglu, A., Davies, S., Smith, P. M., Taylor, P., Simpson, R. J., Ward, L. D., McCluskey, J., Puy, R., and Knox, R. B. 1991. Isolation of cDNA encoding a newly identified major allergenic protein of rye-grass pollen: intracellular targeting to the amyloplast. *Proc. Natl. Acad. Sci. USA* **88,** 1384–1388.

Singh, M. B., Smith, P. M., and Knox, R. B. 1990. Molecular biology of rye-grass pollen allergens. *In* "Molecular Approaches to the Study of Allergens" (B. A. Baldo, ed.), Vol. 28., pp. 101–120. S. Karger, Basel, Switzerland.

Sinnot. 1990.

Singletary, G. W., Banisadr, R., and Keeling, P. L. 1994. Heat stress during grain filling in maize: Effects on carbohydrate storage and metabolism. Austr. *J. Plant Physiol,* **21,** 829–841.

Sivak, M. N., 1992. (1→4)-α-D-Glucan synthesis by a chloroplastic phosphorylase isolated from spinach leaves is independent of added primer. *Carbohyd. Res.* **227,** 241–255.

Sivak, M. N., and Preiss, J. 1994. Starch synthesis in seeds. *In* "Seed Development and Germination" (J. Kigel and G. Galili, eds.), pp. 139–168. Marcel Dekker, New York.

Sivak, M. N., and Preiss, J. 1995. Progress in the genetic manipulation of crops aimed at changing starch structure and increasing starch accumulation. *J. Environm. Polymer Degradation.* **3,** 145–152.

Sivak, M. N., Wagner, M., and Preiss, J. 1993. Biochemical evidence for the role of the waxy protein from pea (*Pisum sativum* L.) as a granule-bound starch synthase. *Plant Physiol.* **103,** 1355–1359.

Sivak, M. N., and Walker, D. A. 1986. Photosynthesis in vivo can be limited by phosphate supply. *New Phytol.* **102,** 499–512.

Skoog, F., and Miller, C. O. 1957. Chemical regulation of growth and organ formation in plant tissue cultured *in vitro. Symp. Soc. Exper. Biol.* **11,** 118–130.

Smith, A. M. 1988. Major differences in isoforms of starch-branching enzyme between developing embryos of round and wrinkled-seeded peas (*Pisum sativum* L.). *Planta* **175,** 270–279.

Smith, A. M., Bettey, M., and Bedford, I. D. 1989. Evidence that the *rb* locus alters the starch content of developing pea embryos through an effect on ADPglucose pyrophosphorylase. *Plant Physiol.* **89,** 1279–1284.

Smith, A. M., Denyer, K., and Martin, C. 1997. The synthesis of the starch granule. *Ann. Rev. Plant Physiol. Plant Mol. Biol.* **48,** 67–87.

Smith, A. M., Neuhaus, H. E., and Stitt, M. 1990. The impact of decreased activity of starch-branching enzyme on photosynthetic starch synthesis in leaves of wrinkled-seeded peas. *Planta.* **181,** 310–315.

Smith, P. K., Khron, R. I., Hermanson, G. T., Mallia, A. K., Gartner, F. H., Provenzano, M. D., Fujimoto, E. K., Goeke, N. M., Olson, B. J., and Kenk, D. C. 1985. *Anal. Biochem.* **150,** 76–85.

Smith-White, B., and Preiss, J. 1992. Comparison of proteins of ADP-glucose pyrophosphorylase from diverse sources. *J. Molec. Evolu.* **34,** 449–464.

Smyth, D. A. 1988. Some properties of starch branching enzyme from indica rice endosperm (*Oryza sativa* L.). *Plant Sci.* **57,** 1–8.

Smythe, C., Caudwell, F. B., Ferguson, M., and Cohen, P. 1988. Isolation and structural analysis of a peptide containing the novel tyrosyl–glucosyl linkage in glycogenin. *EMBO J.* **7,** 2681–2686.

Soll, J., Worzer, C., and Buchanan, B. B. 1983. Fructose-2,6-biphosphate and C4 plants. *Adv. Photosynth. Res.* **3,** 145–148.

Sonnewald, U., and Willmitzer, L. 1992. Molecular approaches to sink-source interactions. *Plant Physiol.* **99,** 1267–1270.

Sonnewald, U., Lerchl, J., Zrenner, R., and Frommer, W. 1994. Manipulation of Sink–Source relations in transgenic plant. *Plant Cell Environ.* **17,** 649–658.

Sowokinos, J. R. 1976. Pyrophosphorylase in *Solanum tuberosum.* I. Changes in ADP-glucose and UDP-glucose pyrophosphorylase. *Plant Physiol.* **57,** 63–68.

Sowokinos, J. R., and Preiss, J. 1982. Pyrophosphorylases in *Solanum tuberosum*. III. Purification, structural and catalytic properties of potato tuber ADP-glucose pyrophosphorylase. *Plant Physiol.* **69,** 1459–1466.

Spilatro, S. R., and Preiss, J. 1987. Regulation of starch synthesis in the bundle sheath and mesophyll of *Zea mays* L. Intercellular compartmentalization of enzymes of starch metabolism and the properties of the ADPglucose Pyrophosphorylases. *Plant Physiol.* **83,** 621–627.

St. Pierre, B., and Brisson, N. 1995. Induction of the plastidic starch phosphorylase gene in potato storage sink tissue: Effect of sucrose and evidence for coordinated regulation of phosphorylase and starch biosynthetic genes. *Planta.* **195,** 339–344.

Stark, D. M., Timmerman, K. P., Barry, G. F., Preiss, J., and Kishore, G. M. 1992. Role of ADPglucose pyrophosphorylase in regulating starch levels in plant tissues. *Science* **258,** 287–292.

Steup, M. 1984. Starch degradation. *In* "The Biochemistry of Plants: Vol. 14. Carbohydrates" (J. Preiss, ed.) pp. 255–296. Academic Press, New York.

Steup, M. 1990. Starch degrading enzymes. *In* "Methods in plant biochemistry" (P. M. Dey and J. B. Harborne, eds.), Vol. 3, pp. 103–128. Academic Press, New York.

Steup, M., Robenek, H., and Melkonian, M. 1983. In vitro degradation of starch granules isolated from spinach chloroplasts. *Planta* **158,** 428–436.

Stitt, M. and ap Rees, T. 1980. Carbohydrate breakdown by chloroplasts of *Pisum sativum.* **627,** 131–143.

Stitt, M., and Heldt, H. W. 1985. Control of photosynthetic sucrose synthesis by fructose-2,6 bisphosphate. *Planta* **164,** 179–188.

Stitt, M., Huber, S. C., and Kerr, P. 1987a. Control of photosynthetic sucrose formation. *In* "The Biochemistry of Plants" (M. D. Hatch and N. K. Boardman, eds.), Vol. 10, pp. 327–409. Academic Press, New York.

Stitt, M., Huber, S. C., and Kerr, P. 1987b. Regulation of photosynthetic sucrose synthesis. *In "The Biochemistry of Plants"* (M. D. Hatch and N. K. Boardman, eds.), Vol. 13, pp. 327–409. Academic Press, New York.

Stitt, M., and Steup, M. 1984. Starch and sucrose degradation. *In* "Encyclopedia of Plant Physiology" (R. Douce and D. Day, eds.) Vol. 18, pp. 347–390. Springer-Verlag, Berlin, Germany.

Sullivan, T., and Kaneko, 1995. The maize brittle 1 gene encodes amyloplast membrane polypeptides. *Planta* **196,** 477–484.

Sullivan, T., Strelow, L. I., Illingworth, C. A., Phillips, C. A., and Nelson, O. E. 1991. The maize brittle-1 locus: Molecular characterization based on c-DNA clones isolated using the dSpm-tagged brittle-1-mutable allele. *Plant Cell.* **3,** 1337–1348.

Svensson, B. 1994. Protein engineering in the α-amylase family: Catalytic mechanism, substrate specificity and stability. *Plant Molec. Biol.* **25,** 141–157.

Swinkels, J. J. 1989. Differences between commercial native starches. AVEBE brochure. Foxhol, The Netherlands.

Takata, H., Kuriki, T., Okada, S., Takesada, Y., Iizuka, M., Minamiura, N., and Imanaka, T. 1992. Action of neopullulanase. Neopullulanase catalyzes both hydrolysis and transglycosylation at α-(1,4) and α-(1,6)-glucosydic linkages. *J. Biol. Chem.* **267,** 18447–18452.

Takeda Y., Guan H. P., and Preiss J. 1993. Branching of amylose by the branching isoenzymes of maize endosperm. *Carbohyd. Res.* **240,** 253–263.

Takeda, Y., and Hizukuri, S. 1987. Structures of rice amylopectins with low and high affinities for iodine. *Carbohyd. Res.* **168,** 79–88.

Takeda, Y., Hizukuri, S., Ozono, Y., and Suetake, M. 1983. Actions of porcine pancreatic and *Bacillus subtilis* α-amylase and *Aspergillus* niger glucoamylase on phosphorylated (1→4)-α-D-glucan. *Biochim. Biophys. Acta* **749,** 302–311.

Takeda, Y., Takeda, C., Suzuki, A., and Hizukuri, S. 1993. Structure of the amylopectin of amylomaize. *Carbohyd. Res.* **246,** 273–283.

Tandecarz, J. S., and Cardini, C. E. 1978. A two-step enzymatic formation of a glucoprotein in potato tuber. *Biochim. Biophys. Acta.* **543,** 423–429.

Tandecarz, J. S., and Cardini, C. E. 1979. Effect of 1,5 gluconolactone on the formation of a glucoprotein in potato tuber. *Plant Sci. Letters.* **15,** 151–158.

Tetlow, I. J., Blissett, K. J., and Emes, M. J. 1993. A rapid method for the isolation of purified amyloplasts from wheat endosperm. *Planta* **189,** 597–600

Thorbjørnsen, T., Villand, P., Denyer, K., Olsen, O.-A., and Smith, A. M. 1996. Distinct isoforms of ADPglucose pyrophosphorylase occur inside and outside the amyloplasts in barley endosperm. *Plant J.* **10,** 243–250.

Tong, L. A., de Vos, A. M., Milburn, M. V., and Kim, S. H. 1991. Crystal structures at 2.2 Å resolution of the catalytic domains of normal *ras* protein and an oncogenic mutant complexed with GDP. *J. Mol. Biol.* **217,** 503–516.

Tsai, C. Y. 1974. The function of the *waxy* locus in starch synthesis in maize endosperm. *Biochem. Genet.* **11,** 83–96.

Tsai, C. Y., and Nelson, O. E. 1966. Starch-deficient maize mutant lacking adenosine diphosphate glucose pyrophosphorylase activity. *Science* **151,** 341–343.

Turner, J. F. 1969. Physiology of pea fruits. VI. Changes in uridine diphosphate glucose pyrophosphorylase and adenosine diphosphate glucose pyrophosphorylase in the developing seed. *Aust. J. Biol. Sci.* **22,** 1321–1327.

Tyson, R. H., and ap Rees, T. 1988. Starch synthesis by isolated amyloplasts from wheat endosperm. *Planta* **175,** 33–38.

Van den Koornhuyse, N., Libessart, N., Delrue, B., Zabawinski, C., Decq, A., Iglesias, A., Carton, A., Preiss, J., and Ball, S. 1996. Control of starch composition and structure supply in the monocellular alga *Chamydomonas reinhardii. J. Biol. Chem.* **271,** 16281–16287.

van der Leij, F. R., Visser, R. G. F., Ponstein, A. S., Jacobsen, E., and Feenstra, W. J. 1991. Sequence of the structural gene for granule-bound starch synthase of potato (*Solanum tuberosum* L.) and evidence of a single point deletion in the *amf* allele. *Mol. Gen. Genet.* **228,** 240–248.

Villand, P., Olsen, O.-A., Kilian, A., and Kleczkowski, L. A. 1992. ADP-glucose pyrophosphorylase large subunit cDNA from barley endosperm. *Plant Physiol.* **100,** 1617–1618.

Villand, P., and Kleczkowski, L. A. 1994. Is there an alternative pathway for starch biosynthesis in cereal seeds? *Z. Naturforsch* **49c,** 215–219.

Viola, R., Davies, H. V., and Chudeck, A. R. 1991. Pathways of starch and sucrose biosynthesis in developing tubers of potato (*Solanum tuberosum* L.) and seeds of faba bean (*Vicia faba* L.): Elucidation by ^{13}C-nuclear magnetic-resonance spectroscopy. *Planta* **183,** 202–208.

Visser, R. G. F., Somhorst, I., Kuipers, G. J., Ruys, N. J., Feenstra, W. J., and Jacobsen, E. 1991. Inhibition of the gene for granule-bound starch synthase in potato by antisense constructs. *Mol. Gen. Genet.* **225,** 289–296.

Vos-Scheperkeuter, G. H., De Boer, W., Visser, R. G. F., Feenstra, W. J., and Witholt, B. 1986. Identification of granule-bound starch synthase in potato tubers. *Plant Physiol.* **82,** 411–416.

Vos-Scheperkeuter, G. H., de Wit, J. G., Ponstein, A. S., Feenstra, W. J., and Witholt, B. 1989. Immunological comparison of the starch branching enzymes from potato tubers and maize kernels. *Plant Physiol.* **90,** 75–84.

Walden, R., Koncz, C., and Schell, J. 1990. The use of gene vectors in plant molecular biology. *Methods Molec. Cell Biol.* **1,** 175–194.

Walker, D. A. 1971. Chloroplasts (and grana): Aqueous (including high carbon fixation ability). *Meth. Enzymol.* **5,** 211–220.

Weber, H., Heim, U., Borisjuk, L., and Wobus, U. 1995. Cell-type specific, coordinate expression of two ADPglucose pyrophosphorylase genes in relation to starch biosynthesis during seed development in *Vicia faba* L. *Planta* **195,** 352–361.

Werr, W., Frommer, W. B., Maas, C., and Starlinger, P. 1985. Structure of the sucrose synthase gene on chromosome 9 of *Zea-mays* L. *EMBO J.* **4,** 1373–1380.

Wessler S. R., Baran, G., Varagona, M., and Dellaporta, S. L. 1986. The excision of *Ds* produces *waxy* proteins with a range of enzymatic activities. *EMBO J.* **5,** 2424–2432.

Whatley, J. B. 1978. A suggested cycle of plastid development interrelationships. *New Phytol.* **80,** 489–502.

Whistler, R. L. 1984. *In* "Starch: Chemistry and Technology" (R. L. Whistler, J. N. BeMiller, and E. F. Paschall, eds.), 2nd ed. Academic Press, New York.

Wurzburg, O. B. 1986. "Properties and Uses for Modified Starches." CRC Press, Boca Raton, Florida.

Yamamoto, T. (ed.) 1995. Enzyme chemistry and molecular biology of amylases and related enzymes: CRC Press, Boca Raton, FL.

Yamanouchi, H., and Nakamura, Y. 1992. Organ specificity of isoforms of starch branching enzymes (Q-enzyme) in rice. *Plant Cell Physiol.* **33,** 981–985.

Yang, H., Liu, M. Y., and Romeo, T. 1996. Coordinate genetic regulation of glycogen catabolism and biosynthesis in Escherichia coli via the CSRA gene product. *J. Bacteriol.* **178,** 1012–1017.

Zamski, E., and Schaffer, A. A. 1996. Photoassimilate distribution in plants and crops. Source–sink relationships. Marcel Dekker, Inc. New York. 905 pp.

Ziegler, P. 1995. Carbohydrate degradation during germination. *In* "Seed Development and Germination" (J. Kigel and G. Galili, eds.), pp. 447–476. Marcel Dekker, New York.

Zrenner, R., Wilmitzer, I., and Sonnewald, U. 1993. Analysis of the expression of potato uridinediphosphate glucose pyrophosphorylase and its inhibition by antisense RNA. *Planta* **190,** 247–252.

INDEX

A

ADPGlc, *see* ADPglucose pyrophosphorylase
ADPglucose
  synthesis
    glycogen, 125
    pathway *in vivo,* 34–37
    role, 39–40
ADPglucose pyrophosphorylase
  amyloplasts, 118–119
  amyloplasts transport, 119–120
  antisense, 129–130
  cloning, 59–68
  cold storage and, 136
  genomic DNA, 73–74
  higher plant subunits, 50–51
  hydrophobic cluster analysis, 68–72
  localization, 123–124, 143
  regulation, 46–47, 145–148
  sink strength, 136
  structure-function, 49–50
  substrate binding sites, 51–59
  subunit structure, 47–49
  synthesis
    pathway, 43–46
    regulation, 131–134
    role, 35–38
  transcription, 72–73
Agrobacterium tumefaciens, 126
Allergens, 32
Amylase
  conservation, 102–103
  mechanisms, 159–160
  starch degradation, 153–154
α-Amylase, 157–158
β-Amylase, 157
Amylograph, Brabender, 167
α-Amylolysis, 156
Amylopectin
  branching enzymes, 95
  Chlamydomonas reinhardth, 85–86
  granule bound synthases, 79
  granules, 13, 16, 20–23
  molecule size, 164
  orientation, 28
  SSS in, 76–77
  structure, 23–25
Amyloplasts
  biosynthesis
    analysis, 116–117
    characterization, 115–116
  carbon transport, 119–124
  cell fractionation, 117–119
Amylose
  branching enzymes, 93–95
  Chlamydomonas reinhardth, 85–86
  granules, 13, 16, 20–23
  molecule size, 164
  orientation, 28
  processing, 169–170
Amylotylic enzymes, 135–136
Antisense, 129–130
Apical kernels, 38–39
Arabidopsis thaliana
  branching enzymes, 100
  as starch model, 10
Autocatalysis, 140

B

Baking products, 165
Biosynthesis
  ADPglucose, 34–37
  alternative pathways, 37–38
  amyloplasts, 115–116

Biosynthesis (*continued*)
  CAM plants, 150–152
  complete pathway, 111–113
  enzymes, limiting, 134–135
  glg C and, 131–134
  immunocytochemical studies, 116–117
  leaves, 148
  manipulation, 135–136
  microscopy, 116–117
  rate, 38–39
  studies, 33–34
Birefringence, 27–28
Brabender amylograph, 167
Branching enzymes
  assay, 89, 91–92
  characterization, 89
  C_4 plants, 150
  genes, 95–98
  mode of action, 93–95
  purification, 92–93
  structure-function, 101–103
Branching-linkage assay, 91–92
Brush border, 158
Bt1 gene, 120
Bundle sheath, 150

C

Callus, 128
Calorimetry, differential scanning, 167–168
CAM plants, 150–152
Carbohydrate metabolism, 144–145
Carbon transport, 119–124
Caryopsis
  defination, 1–2
  development, 7
Cassava, 99
Cauliflower mosaic virus S25, 128
Cell fractionation, 117–119
Cellulose, 158
Chlamydomonas reinhardth
  mutants, 85–87
  as starch model, 11
Chloroplasts
  ADPGlc PPase, 145–148
  characterization, 115
  phosphoglucose isomerase mutants, 146
  starch degradation outside, 156–157
  as transporting organelle, 143–144
C_4 plants, 148–150
Crosslinking, 168
Cyanobacteria, 6

D

Debranching enzymes
  degradation, 154–155
  granule formation, 110–111
Degradation
  amylase, 153–154
  debranching enzymes, 154–155
  outside chloroplast, 156–157
  pathway, 155–156
  phosphorylase, 153–154
Dextrins, 158, 169
Differential scanning calorimetry, 167–168
Digestion, human, 157–158
Dissociation, 9
DNA
  ADPGlc, 73–74
  exogenous
    characterization, 125–126
    T_i-plasmid, 126–128

E

Escherichia coli, 43

F

Floridean starch, 5–6
Fructans, 5

G

GBSS, *see* Granule-bound starch synthase
Gelatinization, 165–168
Gene disruption, 10
Genetic engineering, 125–126
glg C gene, 132–134
Glucans
  reserve, 5
  starch degradation, 153
Glucose-6-P dihydroxyacetone-P, 120
Glucose 6-phosphate, 30–31
Glucose transport, 159
α-1,4-Glucosidic linkages, 33–34
Glycogen
  characterization, 6
  synthesis, 125

Glycogen synthase, 100
  characterization, 107–108
  sequences, 85
Glycogen-synthesizing enzymes, 110
Grains, *see* Caryopsis
Granule-bound starch synthases
  antisense, 129
  characterization, 78–81, 111–113
Granules
  amylopectin, 13, 16, 20–23
  amylose, 13, 16, 20–23
  dried, 165
  formation, 110–111
  lipids in, 30
  minor constituents, 30
  orientation, 27–29
  phosphorus in, 30
  proteins in, 31–32
  starch synthases bound, 78–81
  structure, 13
Grasses, 2
Green algae, 4

H

HCA, *see* Hydrophobic cluster analysis
Hexose-P, 122
Hydrophobic cluster analysis, 68–72

J

Jumping genes, 7

K

Kernels, *see* Caryopsis

L

Laminarin, 5
Leaves, 3–4, 148
Legumes, 2
Light microscopy, 167
Lintnerized starch, 168
Lipids, 30
Lysine, 53–55

M

Maize
  biosynthesis rate, 38–39, 135
  branching enzymes in, 92–93
  enzymes, 47–48
  SSS in, 76–77
  as starch model, 7–9
  starch synthases, 80–81
  waxy gene, 82
Male sterility, 131
Mannans, 5
Mesophyll, 150
Metabolic pathway
  altering, 125–126
  vectors, 126–127
Metabolism
  Arabidopsis thaliana model, 10
  carbohydrate, 144–145
  Chlamydomonas reinhardth model, 11
  maize model, 7–9
  model comparisons, 6–7
  potato model, 9–10
  snapdragon model, 10–11
  wheat model, 6–7
Microscopy, 167
Microvilli, 158
Mouth-feel, 165

N

Naegeli starch, 168
Neutralization, 95–96
Nomenclature, 29

O

Organelles
  expression specific to, 128–129
  transporting, 143–144

P

Pancreatic amylase, 158
Pasting behavior, 165–166
Peas
  branching enzymes, 100–101
  SSS in, 84–85
Phosphoglucose isomerase mutants, 146
3-Phospho-glycerate, 145–148
Phosphorus starch granules, 30
Phosphorylase
  mechanisms, 159–160
  starch degradation, 135–136, 153–154

Phosphorylase-stimulation assay, 89
Photosynthesis cells, 139–140
Plants, *see also specific species*
  C_4, 148–150
  regeneration, 128
  starch degradation pathway, 155–156
  transformation, 131–134
Polarimetry, 167
Polysaccharides, *see also specific types*
  degradation, 160
  reserve, 4–6
Porin, 144
Potatoes
  ADPGlc PPase, 48–49
  branching enzymes, 98–99
  branching enzymes in, 93
  characteristics, 164
  life cycle, 10
  modification, 129
  starch in, 3
  as starch model, 9–10
Prepared foods, 165
Promoters, 128–129
Proteins
  expression, 128–129
  starch granules, 31–32
Protoplast transformation, 127–128
Pyridoxal phosphate, 53–55
Pyrophosphatase, 130

R

*rbc*S gene, 132
Reductive pentose phosphate pathway, 140–141, 143
Regeneration, 128
Reserve glucans, 5
Reserve polysaccharides, 4–6
Retrogradation, 20, 166
r gene, 100–101
Rhodobacter spheroides, 44
Rhodospirillum rubrum, 43
Rice
  branching enzymes, 99–100
  SSS in, 78
Roots, 3
RPPP, *see* Reductive pentose phosphate pathway
rr gene, 134

S

Salivary α-amylase, 158
Seeds, 1–2
Shear thinning, 165
Snapdragons, 10–11
Soluble starch synthases
  characterization, 76
  classification, 80–81
  forms, 75–78
  granule bound, *see* Granule-bound starch synthase
  starch biosynthesis, 111–113
  waxy gene, 81–82, 84–85
Spinach leaf enzyme, 47–48
SSS, *see* Soluble starch synthase
Starch
  biosynthesis
    ADPglucose, 34–37
    alternative pathways, 37–38
    CAM plants, 150–152
    complete pathway, 111–113
    C_4 plants, 148–150
    enzymes, limiting, 134–135
    glg C and, 131–134
    initiation, 107–110
    leaves, 148
    manipulation, 135–136
    rate, 38–39
    studies, 33–34
  caryposis, 1–2
  characterization, 1
  degradation, 135–136
    amylase, 153–154
    debranching enzymes, 154–155
    outside chloroplast, 156–157
    pathway, 155–156
    phosphorylase, 153–154
  dextrins, 158
  digestion, human, 157–158
  granules
    amylopectin, 13, 16, 20–23
    amylose, 13, 16, 20–23
    dried, 165
    formation, 110–111
    lipids in, 30
    minor constituents, 30
    orientation, 27–29
    phosphorus in, 30
    proteins in, 31–32
    structure, 13

green algae, 4
industrial
analysis, 167–168
applications, 163–164
leaves, 3–4
lintnerized, 168
manufacturing, 164–166
metabolism
Arabidopsis thaliana model, 10
Chlamydomonas reinhardth model, 11
maize model, 7–9
model comparisons, 6–7
potato model, 9–10
snapdragon model, 10–11
wheat model, 6–7
methodology, 29
modification, 168–169
molecules, crosslinking, 168
Naegeli, 168
nomenclature, 29
photosynthesis cells, 139–140
polymers, biodegradable, 169–170
properties, 164–166
roots, 3
seeds, 1–2
soluble, *see* Soluble starch
storage tissues, 117
sweetener conversion, 169
synthesis
genetic engineering, 125–126, 130–131
vectors, 126–127
tubers, 3
Sterility, 131
Storage substance, 4
Storage tissues, 117
Sucrose synthesis, 148–150
Sugary 1
characterization, 110–111
debranching, 154–155
effects, 40
Sweetener conversion, 169
Swimming cells, 11
Syneresis, 166
Syrup production, 163–164

T

TCA, *see* Trichloroacetic acid
T_i-plasmid, 126–128
Tissue-specific expression, 128–129
Transgene, 126
Transporting organelles, 143–144
Trichloroacetic acid, 108
Triose-P isomerase, 120, 122
Triticale, 38
Tubers, 3

U

UDP-glucose
regulation, 125
role, 35–36, 39–40
UDPglucose:protein transglucosylase, 109–110
UPTG, *see* UDPglucose:protein transglucosylase

V

Vegetative cells, 11
Viscosity, 165, 167

W

Waxy gene, 81–82, 84–85
Wheat
breeding, 131
characterization, 6–7
gene families, 59
glycemic response, 159
modification, 130–131
starch synthesis, 121

X

X-ray analysis cap, 167